Advanced Elastomeric Materials

ADVANCED ELASTOMERIC MATERIALS

Edited by:

Saeed Farrokhpay

ARCLER PRESS

www.arclerpress.com

Advanced Elastomeric Materials

Saeed Farrokhpay

Arcler Press
224 Shoreacres Road
Burlington, ON L7L 2H2
Canada
www.arclerpress.com
Email: orders@arclereducation.com

ISBN: 978-1-77469-074-1 (Hardcover)

ABOUT THE EDITOR

Dr Saeed Farrokhpay is a Chemical Engineer with several years of experience in mineral & material processing. He obtained his PhD from University of South Australia in 2005. He is currently a Technical Consultant in Australia. He has worked for more than 20 years at mineral and chemical industries, universities and research centers around the world. Dr Farrokhpay has published more than 90 papers in high ranked journals and conference proceedings. He has also edited several technical and scientific books, and served as an editorial board member of several international scientific journals.

TABLE OF CONTENTS

List of Figures ..... *xi*
List of Tables ..... *xvii*
List of Abbreviations ..... *xix*
Preface ..... *xxiii*

Chapter 1 Introduction to Elastomeric Materials 1
1.1. Introduction 2
1.2. Thermodynamics 5
1.3. Kinetics 11
1.4. Structure 12
1.5. Polymerization 14
1.6. Particular Elastomer Structure and its Properties 14
1.7. Thermoplastic Elastomer (TPE) Structure and Properties 28
1.8. Elastomer-Filler Compositions 33
1.9. Elastomer Blends 35
1.10. Interpenetrating Elastomer Blends (IPN) 36
1.11. Formulation And Compounding 37
1.12. Shape-Memory Polymers 37
1.13. Summary and the Future Trends 38
References 40

Chapter 2 Classification of Elastomeric Materials 51
2.1. Introduction 52
2.2. Natural Rubber (NR) 55
2.3. Isoprene Rubber, Polyisoprene (IR) 58
2.4. Butadiene Rubber (BR), Polybutadiene 60
2.5. Styrene-Butadiene Rubber (SBR) 63
2.6. Butyl Rubbers 67

2.7. Nitrile Rubber, Nitrile-Butadiene Rubber, and Acrylonitrile Rubber (NBR) 69
2.8. Epichlorohydrin Rubbers 71
2.9. Ethylene-Propylene Rubbers 73
2.10. Chloroprene Rubber, Polychloroprene (CR) 76
2.11. Polyacrylate Rubbers (ACM) 77
2.12. Polyurethane Rubbers (AU, EU, PUR) 79
2.13. Fluorocarbon Rubbers (FKM, FPM) 82
2.14. Silicone Rubbers (Q) 85
2.15. Polysulfide Rubbers (T) 87
2.16. Ethylene-Vinyl Acetate Copolymer (EVA) 89
2.17. Polypropylene Oxide Rubbers, Polypropylene Oxide-Allyl Glycidyl Ether Copolymer (GPO) 90
2.18. Chlorinated Polyethylene (CM, CPE), Chlorosulfonated Polyethylene (CSM, CSPE) 91
2.19. Thermoplastic Elastomers (TPE) 93
2.20. Styrenic Thermoplastic Elastomers (TPE-S) 93
2.21. Elastomeric Alloys 95
2.22. Thermoplastic Urethane Elastomers (TPU, TPE-U) 97
2.23. Thermoplastics Polyester-Ether Elastomer (TPE-E) 99
2.24. Thermoplastic Polyamide Elastomers (TPE-A) 100
References 103

Chapter 3 Elastomeric Blends 115
3.1. Introduction 116
3.2. Thermodynamics of Blends of the Rubber 117
3.3. Thermodynamics Aspects Incorporating the Filler Effect in the Rubber Blends 120
3.4. Miscible Rubber Blends 123
3.5. Immiscible Rubber Blends 124
3.6. Compatible Rubber Blends 124
3.7. Incompatible Rubber Blends 125
3.8. Factors Affecting the Performance of Blends of Rubber 125
3.9. Distribution of the Additives of Rubber In Rubber Blends 127
3.10. Preparation of the Rubber Blends 129
3.11. Correlation Between Viscosity and Polarity 132

3.12. Morphology Categorization by Microscopy 133
3.13. Preparation Methods Utilized for Microscopy 133
3.14. Om (Optical Microscopy) of the Rubber Blends 136
3.15. Morphology of the Rubber Blends Through Sem 140
References 146

Chapter 4 **Mechanical and Physical Characteristics of Rubber Compounds 159**
4.1. Introduction 160
4.2. Fatigue of Rubber 160
4.3. Crack Growth Examination and Fracture 163
4.4. Rubber Filler Interaction 166
4.5. Rubber Crystallization 168
4.6. Rubber Molecular Orientation 170
References 171

Chapter 5 **Conductive Elastomers and Their Electronic Applications 177**
5.1. Introduction 178
5.2. Stretchable Electronics 180
5.3. Stretchable Sensors 188
5.4. Stretchable Energy Harvesters 194
5.5. Summary 198
References 199

Chapter 6 **Mechanical and Electrical Characteristics of the Elastomeric Nanocomposites 207**
6.1. Introduction 208
6.2. Basic Problems on Carbon Nanomaterials 209
6.3. Manufacturing Methods of Rubber Nanocomposites 210
6.4. Tensile Properties 211
6.5. Dynamic Mechanical Properties 213
6.6. Electrical Properties Under Strain 217
References 219

Chapter 7 **Applications of Elastomers in Defense and Aerospace 223**
7.1. Introduction 224
7.2. Acoustic Applications 224
7.3. Aircraft Tires 230

7.4. Airships 234
7.5. Inflatable Seacraft 237
References 240

Chapter 8 Bio-Medical Uses of Elastomeric Materials 245
8.1. Introduction 246
8.2. Developments on Elastomeric Blends and Composites for Medical Applications 247
8.3. Biostable and Biocompatible Materials 249
8.4. Modern Elastomer's Superior Mechanical Properties 249
8.5. Intelligent Bio-Materials 250
8.6. Overview of Products Based on Elastomer Biomedical Applications 250
8.7. Thermoplastic Composites and Elastomers 251
References 259

Index 263

LIST OF FIGURES

Figure 1.1. The elastic stress versus strain curves of the thermoplastic matched with an elastomer

Figure 1.2. A poly(cis-1,4-isoprene) random coil having 15 monomer units

Figure 1.3. Two cross-linked 15 mers of a poly(cis-1,4-isoprene)

Figure 1.4. Temperature, length, and volume associations for an elastomer and gas

Figure 1.5. Viscoelastic, viscous, and elastic component models

Figure 1.6. Stress versus strain time (upper) and the hysteresis (lower) curves for a poly(cis-1,4-isoprene) elastomer

Figure 1.7. Stress versus strain time (upper) and the hysteresis (lower) curves for the poly(styrene-co-butadiene) elastomer

Figure 1.8. Creep strain (creep dotted) and the recovery strain (recovery dotted) response for the poly(cis-1,4-isoprene) elastomer under the ductile creep stress of nearly 0.5 MPa trailed by recovery with the least restraining force

Figure 1.9. Creep strain (creep dotted) and the recovery strain (recovery dotted) response for the poly(butadiene-co-styrene) elastomer under the ductile creep stress of almost 1.0 MPa trailed by the least restraining force

Figure 1.10. A graphic demonstration of the hydrocarbon elastomer properties (4 excellent; 3 good; 2 fair; 1 poor)

Figure 1.11. Property contrast of chloroprene and nitrile elastomers (5 excellent; 4 very good; 3 good; 2 fair; 1 unsatisfactory). The ratings are dependent on compound composition; hence all of the optimal values might not be attained at the same time)

Figure 1.12. Stress versus strain time and the hysteresis curves for the poly(ether-co-urethane) thermoplastic elastomer

Figure 1.13. Chemical resistance of the elastomers compared with fluorocarbon elastomers

Figure 1.14. Relative market segments of the fluorocarbon elastomers displaying specialty automotive supremacy

Figure 1.15. World TPE demand in 2014, 2009, and 2004 by types of thermoplastic elastomers and by regions

Figure 1.16. Diagram of the (styrene-diene-styrene) block copolymer

Figure 1.17. Growth of morphology in the thermoplastic vulcanizates from the co-continuous phase to isolated phase

Figure 1.18. Graphic structure for the interpenetrating polymer blend

Figure 2.1. Two-dimensional chemical structure of4-polyisoprene, cis-1

Figure 2.2. Gathering of the latex from the rubber tree

Figure 2.3. Techniques of processing latex into commercial grades of the dry natural rubber

Figure 2.4. The isomeric structures of polyisoprene

Figure 2.5. The isomeric structures of BR

Figure 2.6. The structure of styrene and isomeric structures of polybutadiene

Figure 2.7. Butyl rubber synthesized from isoprene units and isobutylene

Figure 2.8. Acrylonitrile and butadiene units

Figure 2.9. The polymer made from the structures of ECO and CO

Figure 2.10. The structure of EPM rubbers

Figure 2.11. Isomeric structures of CR

Figure 2.12. The simple structure of acrylates

Figure 2.13. Instances of the structure of acrylate rubbers

Figure 2.14. Creating of urethane group

Figure 2.15. Typical polyols utilized in polyurethanes

Figure 2.16. Typical diisocyanates that are utilized in polyurethanes

Figure 2.17. An instance of a structure of the fluorocarbon rubbers, VF_2/HPTFP/TFE copolymer

Figure 2.18. The structure of silicone

Figure 2.19. The polymerization of polysulfide. Reactants are sodium sulfide and ethyl chloride

Figure 2.20. Ethylene-vinyl acetate rubber

Figure 2.21. Polypropylene oxide rubber

Figure 2.22. Chlorinated polyethylene

Figure 2.23. Chlorosulfonated polyethylene

Figure 2.24. The radial and linear structure of styrene thermoplastic

Figure 2.25. The structure of TPE-V, displaying finely dispersed vulcanized rubber particles in the thermoplastics matrix

Figure 2.26. The impact of rubber particle size in TPE-V (AES)

Figure 2.27. Creating of urethane group

Figure 2.28. The structure of thermoplastic urethane elastomers: Ether diol chains or long ester and hard urethane segments

Figure 2.29. The structure of thermoplastic polyamide elastomers

Figure 3.1. Structure and morphology in rubber blends

Figure 3.2. Phase image for the blend of rubber, displaying: (a) UCST; (b) LCST; and (c) UCST þ LCST

Figure 3.3. Cataloging chart of the rubber blends

Figure 3.4. Anticipated representation for the miscible rubber blends

Figure 3.5. Schematic representation for the immiscible blends

Figure 3.6. Schematic diagram of (a) incompatible blends of rubber-rubber; and (b) compatible blends of rubber-rubber

Figure 3.7. Pictorial diagram of the cross-linked usual blend of rubber with additives

Figure 3.8. Preparation of sample for microscopy analysis

Figure 3.9. Optical images of natural rubber/styrene-butadiene rubber blends with (a) ionic liquid; and (b) without ionic liquid

Figure 3.10. Phase morphology of the 50 SSBR/50 trans polyisoprene binary blends hardened at 80°C for (a) 0 hours, (b) 20 hours, (c) 60 hours, and (d) 140 hours

Figure 3.11. Schematic representation of continuous and dispersed phases in the rubber-rubber blends

Figure 4.1. Graphic representation of crack propagation in rubbers

Figure 4.2. Rubber-filler collaborations

Figure 4.3. Differential skimming calorimetry graph of the general rubber compound

Figure 5.1. Optical images of (a) a 3D PDMS (poly(dimethylsiloxane) film on PDMS support; and (b) a folded 3D PDMS film. Scale bar, 1 cm; (c) top view SEM image of net-made 3D PDMS film. Scale bar, 1 μm; (d) conductivity of 3D PDMS-EGaln (eutectic gallium indium) stretchable conductor below strains of up to 220%; (e) conductivity variation relying on the number of stretching-releasing cycles under diverse strains

Figure 5.2. (a) Schematic picture; (b) focused SEM image; and (c) low-magnification SEM image of a high stretchable transistor comprising completely of stretchable components; (d) photo images of the stretchable transistor range at two strain states ($\varepsilon = 0$ and 0.7); ID-VG curves relying on (e) the applied strain and (f) the number of stretching release cycles at $\varepsilon = 0.7$. In (a) and (c), D, G, and S represent drain, gate, and source correspondingly

Figure 5.3. (a) Top-view SEM image of the crumpled organic memory; (b) current-voltage characteristics displaying a memory behavior; strain-dependent (c) memory switching; and (d) retention time

Figure 5.4. (a) SEM image of GO (graphene oxide)-soldered AgNW (silver nanowire) junctions. Red arrows specify GO parts wrapping about AgNW junctions; (b) schematic drawing of a stretchable PLED structure; (c) optical photographs of a PLED (polymer light-emitting device) operating at 14 V under diverse strains

Figure 5.5. (a) Cross-sectional SEM picture of PDMS-AgNWs-PDMS sandwich-structured strain sensor. Repeated replies of the strain sensor (b) to stretching releasing cycles at a strain of 70%; and (c) to bending cycles in the bending angles of 10°–90°; (d) presentation of finger motion detection utilizing the strain sensor

Figure 5.6. (a) Schematic image of a micro-pyramid PDMS array. Discrete PDMS pyramids are covered with a PEDOT: PSS-PUD (Polyurethane dispersion) mixture, which assists as a piezoresistive electrode; (b) finite element analysis data displaying stress distributions and SEM pictures at diverse magnitudes of pressures; (c) relative existing changes relying on the applied pressure; however, the sensor is stretched to a specific elongation. Here, LS1 and LS2 signify the linear sensitivities in particular regions

Figure 5.7. (a) Schematic drawing of a whole elastomeric gated temperature sensor; (b) reply of the temperature sensor after cyclic stretching of 0–10,000 cycles at a strain of 30%

Figure 5.8. (a) Diagram of the ultrathin organic solar cell; (b) stretchable solar cell made simply through attaching the ultrathin cell to a pre-stretched elastomer. It could be re-stretched and compressed; (c) current-voltage curves of the solar cell relying on the compression in the range of 0%–80%. Between the black (0%) and purple (80%) lines, individual colors signify the compression growing with a step of 10%. The black dashed line signifies the device after being restored to its early state

Figure 5.9. (a) Diagram picture of the hyper-stretchable nanocomposite generator (SEG); (b) the SEG could be stretched and free without harm; (c) generated (i) open-circuit voltage and (ii) short-circuit current relying on periodic stretching releasing cycles at a strain of 200%

Figure 5.10. (a) Schematic image for the preparation processes of PPY-SWCNT nanocomposites; (b) reliance on the thermoelectric performance of the composite on the mechanical stretching. The inset is an SEM picture of the composite film after a 2.5% stretching

Figure 6.1. Styrene-butadiene rubber (SBR) filled with the double filling (5 phr of MWCNTs + 5 phr of carbon black) (Bokobza et al., 2015)

Figure 6.2. Stress versus strain curves of styrene-butadiene rubber; (a) and natural rubber; (b) composites filled with CB and MWCNTs (in red)

Figure 6.3. Strain dependency of storage modulus of the {poly(dimethylsiloxane)} filled with several amounts of particles of the silica produced in situ by the process of sol-gel

Figure 6.4. Strain dependency of storage modulus: at 25°C of natural rubber composites filled with MWCNTs and CB (a); at 25°C of natural rubber composites filled with carbon nanotubes (5 phr) and with decreased GO (2.5 phr) (CG600 and CG200) as well as with carbon nanotubes (2.5 phr) (b)

Figure 6.5. Strain dependency of electrical resistivity for the NR filled with three phr of MWCNTs. Dots in red color: resistivity during the process of retraction. The other colors signify the different cycles of loading

Figure 7.1. Sonar rubber bow dome

Figure 7.2. After a terrorist attack in Yemen in October 2000, the guided-missile vandal USS Cole back to the United States

Figure 7.3. [Left image]: Bow of the Zr. Ms. Mercuur (A900); [right image] interior sonar dome structure

Figure 7.4. Composite keel dome consisting of a soft rubber core and two resin-impregnated fiberglass outer layers. The geometry of the system is adjusted to be acoustically transparent over a restricted range of frequency

Figure 7.5. Thrust reducer utilized for decoupling the submarine propeller shaft. The natural rubber is intervening between an outer layer of polyethylene and the steel ring

Figure 7.6. Military automobile with polyurea-coated steel plates on back and sides of the truck and sides of HHMWV

Figure 7.7. Fragment-simulating 0.50 caliber projectile striking steel armor plate covered with 1,4-polybutadiene rubber

Figure 7.8. Military crew mounts a tire of a B-52H Stratofortress. The changed tire is drained after its removal. The replacement process takes approximately 20 minutes

Figure 7.9. The world's largest warship, Nimitz-class aircraft carrier. Maneuvering needed to locate planes for launching is subjected to much of the tread wear

Figure 7.10. Trial of various Naval airships (in circa 1930), including a kite balloon (upper craft showing in the figure at left), 5 spherical balloons, two J-class non-rigid blimps in the center, the USS Los Angeles in the middle distance, and a ZMC-2, a rigid metal-skinned blimp (showing at right)

Figure 7.11. Artist's vision of huge blimp WALRUS proposed for swift strategic airlifts

Figure 7.12. Prototype space blimp being float tested, 53 m long

Figure 7.13. Combat rubber raiding craft carrying military personnel between the shore and ship

Figure 7.14. Landing craft air cushion with variable-pitch propellers, and propelled by two four-bladed 3.7 m diameter. The polychloroprene/nylon tube can be viewed at the water interface

Figure 8.1. Polyurethane polymerization (showed on bottom) from a hydroxyl group (showed on top right) and diisocyanate (showed on the top left)

Figure 8.2. At the top, it is an example of artificial heart valves: at the bottom left, the heart demonstrating heart valves positions: i.e., mechanical heart valve is shown, and at the bottom right: polymeric heart valve is shown

Figure 8.3. Polyethylene polymerization from its ethylene, monomer

Figure 8.4. Hip joint replacement prosthesis' schematic

Figure 8.5. Silicone synthesis (shown at bottom) from its monomer (called polydimethylsiloxane)

Figure 8.6. A finger prosthesis implant at left and at right shows when it is implanted on hand

LIST OF TABLES

Table 1.1. Comparison of the key hydrocarbon elastomers (4 excellent; 3 good; 2 fair; 1 poor)

Table 1.2. The contrast of the mechanical properties of 2 polysulfide elastomers

Table 1.3. Structure and abbreviation of elastomers

Table 1.4. Chemical and mechanical properties of elastomers

Table 2.1. Polybutadiene rubbers according to the catalyst utilized

Table 2.2. Comparison the properties of NR and BR

Table 2.3. Properties of solvent polymerization and emulsion polymerization

Table 2.4. Comparison amongst the properties of NR and SBR

Table 2.5. Monomer contents that are usual of tire tread blends

Table 2.6. The influence of monomer content and Tg:n on the properties of tires

Table 2.7. Properties of ethylene-propylene rubbers

Table 2.8. The simple monomers of acrylate rubbers

Table 2.9. Monomers utilized in fluorocarbon rubbers

Table 2.10. Structures of fluorocarbon rubbers

Table 2.11. Silicone rubbers and their pendant group structure

Table 2.12. The properties of TPEU and TPAU

Table 2.13. Comparison of different TPEs

Table 3.1. Factors affecting the performance of blends of the rubber

Table 3.2. Different methods of preparation for rubber blends

Table 5.1. Representative accomplishments made in the field of stretchable electronics

Table 5.2. Representative successes made in the field of stretchable sensors

Table 8.1. World's annual biomaterials consumption (approximated)

LIST OF ABBREVIATIONS

1D-1R	one diode-one resistor
ADS	air-dried sheets
AFM	atomic force microscopy
Ag	silver
AuNP	gold nanoparticle
BIIR	bromobutyl rubber
BR	butadiene rubber
C_2H_5OH	ethanol
CB	carbon black
CIIR	chlorobutyl rubber
CNTs	carbon nanotubes
CO	copolymer
COPAs	polyamide-elastomer block copolymers
COPEs	polyether ester-elastomer block copolymers
CPE	chlorinated polyethylene
CR	chloroprene rubber
CRRC	combat rubber raiding craft
CSPE	chlorosulfonated polyethylene
CVD	chemical vapor deposition
ECO	ethylene oxide copolymer
EPDM	ethylene propylene diene monomer
EPDM-g-MAH	EPDM grafted maleic anhydride
EPM	ethylene-propylene rubber
EQE	external quantum efficiency
ETER	epichlorohydrin terpolymer
EU	urethane rubber
EVA	ethylene-vinyl acetate

FdL	length and force change
FKM	fluorocarbon elastomers
FWHM	full width half maximum
GNPs	graphite nanoplatelets
GO	graphene oxide
HMPE	high-modulus polyethylene
HNBR	hydrogenated nitrile butadiene rubber
HPPE	high-performance polyethylene
HTL	hole transport layer
I_2	iodine
IIR	isobutylene-isoprene rubber
IL	ionic liquid
IPN	interpenetrating polymer network
iPP	isotactic polypropylene
ITO	indium tin oxide
K	potassium
LCAC	landing craft air cushion
LCST	lower critical solution temperature
LEDs	light-emitting diodes
MPR	melt-processible rubbers
mTDSC	modulated temperature differential scanning calorimetry
MWCNTs	multi-walled CNTs
NBR	acrylonitrile-butadiene rubber
NG	nanogenerator
NR	natural rubber
O_2	oxygen
OENR	oil-extended natural rubber
OM	optical microscopy
OPV	organic photovoltaic
P3BT	poly(3-butylthiophene)
PA	polyacetylene
PANI	polyaniline
PCE	power conversion efficiency
PCEA	polycarbonate-ester amide

PDMS	polydimethylsiloxane
PDMS-b-PEO	poly(dimethylsiloxane-*b*-ethylene oxide)
PdV	volume and pressure change
PE	polyethylene
PEA	polyesteramide
PE-b-A	polyether-block-amide
PEC	piezoelectric elastic composite
PEDOT	poly(3,4-ethylene dioxythiophene)
PEEA	polyetheresteramide
PEI	polyethyleneimine
PET	poly(ethylene terephthalate)
PMMA	poly(methyl methacrylate)
PMN-PT	lead magnesium niobate-lead titanate
POSS	polyhedral oligomeric silsesquioxanes
PPY	polypyrrole
PSR	polysulfide rubber
Pt	platinum
PU	polyurethane
PUA	poly(urethane acrylate)
PUD	PU dispersion
PUR	polyurethane rubbers
PVC	polyvinyl chloride
PVDF	polyvinylidene fluoride
RSS	ribbed smoked sheets
RTD	resistance temperature detector
$SbCl_5$	antimony pentachloride
SBCs	styrene block copolymers
SBS	styrene-butadiene-styrene
SEBS	styrene/ethylene-butylene copolymer
SEEPS	styrene ethylene-ethylene/propylene styrene
SEM	scanning electron microscopy
SEPS	styrene/ethylene propylene copolymer
SIS	styrene-isoprene-styrene
SR	silicone rubber

SWCNTs	single-walled CNTs
TEGs	thermoelectric generators
TEM	transmission electron microscopy
TESPT	triethoxy silyl propyl tetrasulfide
THF	tetrahydrofuran
TPE	thermoplastic elastomer
TPE-A	thermoplastic polyamide elastomers
TPE-E	thermoplastics polyester-ether elastomer
TPI	trans polyisoprene
TPO	thermoplastic olefin elastomers
TPU	thermoplastic urethane
TPVs	thermoplastic vulcanizates
UCST	upper critical solution temperature
VLNP	very long nanowire percolation
XNBR	carboxylated nitrile rubbers

PREFACE

Elastomeric materials are extensively in many engineering and domestic applications. Elastomeric materials are in high demand due to their widespread use in extreme conditions, which include corrosive environments and high temperatures. These escalating demands are making it difficult to predict the expected service life of the materials to guarantee their long-term performance. This book thoroughly reviews the wide-ranging research on understanding the fundamentals of elastomeric materials and their manufacturing techniques, along with an emphasis on their properties.

Elastomeric materials are a category of polymers with a highly flexible nature. These materials possess a low level of intermolecular forces that permit them to experience high degrees of extension (elongation). Elastic modulus of elastomeric materials is usually low, and they often remain in an amorphous state. Some crystallinity (in the structure) can also be observed during the stretching of the elastomers. The cross-linking in elastomers imparts the ability to recover their original shape after the external stress (either static or dynamic) is removed. Elastomers usually have a glass transition temperature (Tg) lower than the standard room temperature.

This book contains eight chapters which mainly focus on the fundamentals of various aspects of elastomers. Chapter 1 deals with the introduction of elastomeric materials. Detailed discussion about thermodynamics, kinetics, polymerization, and elastomeric materials is contained in the chapter. Chapter 2 focuses on the classification of elastomeric materials. It deals with the discussion about natural rubber, synthetic rubber, and different types of rubbers. Chapter 3 contains information about different types of elastomer blends. Morphology and compatibility of different elastomeric blends are discussed in the chapter.

Chapter 4 contains the essentials of mechanical properties of different elastomeric materials and their response under mechanical loading. Chapter 5 focuses on the properties and applications of conductive elastomeric materials. Stretchable electronic devices manufactured from elastomeric materials are also discussed in the chapter. Presently, composite materials are gaining significant attention due to their widespread applications. Moreover, the advent of nanotechnology has also revolutionized the industrial sectors. Chapter 6 discusses the mechanical and electrical properties of elastomeric nanocomposites.

Aerospace and defense sectors extensively use elastomeric materials. The developments in elastomeric materials are greatly benefiting these industries. Chapter 7 deals with the description of applications of elastomers in the aerospace and defense sectors. Finally, Chapter 8 deals with biomedical applications of elastomeric materials.

The book is equally beneficial for students, researchers, teachers, and industrialists in the field of elastomers and rubber materials. The book can also be used as a ready reference by undergraduate and post-graduate students. Moreover, people from multidisciplinary fields can also benefit from the book.

CHAPTER 1

Introduction to Elastomeric Materials

Contents

1.1. Introduction ... 2
1.2. Thermodynamics ... 5
1.3. Kinetics ... 11
1.4. Structure ... 12
1.5. Polymerization ... 14
1.6. Particular Elastomer Structure and its Properties ... 14
1.7. Thermoplastic Elastomer (TPE) Structure and Properties ... 28
1.8. Elastomer-Filler Compositions ... 33
1.9. Elastomer Blends ... 35
1.10. Interpenetrating Elastomer Blends (IPN) ... 36
1.11. Formulation And Compounding ... 37
1.12. Shape-Memory Polymers ... 37
1.13. Summary and the Future Trends ... 38
References ... 40

1.1. INTRODUCTION

Elastomers are exclusive to polymers and display astonishing reversible extensions with minimal permanent set and low hysteresis. They are the perfect polymers free of molecular interactions, chain rigidity, and crystallinity constraints. The usual elastomers have distinguishing low modulus, however, with poor chemical resistance and abrasion. Theoretical ideas have been developed for their kinetics and thermodynamics, and this information has been made practical for encompassing their properties by the design of molecular and chemical structures, or through adjustment by control of blending, additions of fillers, or crosslinking (Qi et al., 2003, 2004). This chapter provides a review of elastomer theory and the challenging range of properties anticipated. The natural rubber (NR) is an opening material for the institution of chemistries that present damping, higher modulus, and abrasion resistance through co-polymerization and interrelating functional groups. Heteroatoms like fluorine, oxygen (O_2), nitrogen, and silicon are displayed to encompass properties and provide chemical resistance. The thermoplastic elastomers (TPEs) move away from usually cured systems because of the creation of two-phase block copolymers. Lastly, adjustment by filler and the blended systems is well-thought-out, trailed by an introduction in order to shape the memory materials and a short comment on the trends of the future. The diverse and unique properties and performance of the elastomers remain to be a captivating field for science (Brackbill et al., 1996; Pandey and Mehtra, 2014).

An elastomer is a material that can display a quick and large reversible tension in response to stress. Generally, an elastomer is differentiated from the material that displays an elastic response which is a feature of various materials. The elastic response is a response where the strain, according to Hooke's Law, is proportional to stress, although the strain might only be a small amount, like 0.001 for the silicate glass (Goor et al., 2017).

An elastomer can display a large strain of, for instance, 5 to 10, and to be capable to accomplish this, an elastomer should be a polymer (Flory, 1953). Figure 1.1 displays the comparison stress versus strain curve for the NR matched with a usual thermoplastic, and the brittle plastic. The variation in the strength of the polymers with comparable structures because of the crystallinity of the polypropylene is obvious (Chester and Anand, 2010).

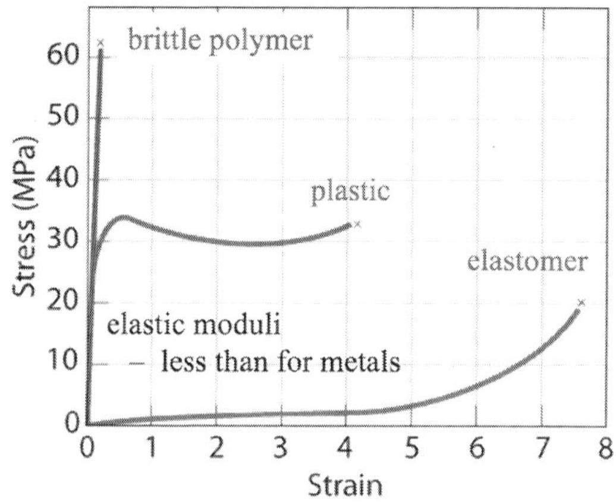

Figure 1.1. The elastic stress versus strain curves of the thermoplastic matched with an elastomer.

Source: https://www.e-education.psu.edu/matse81/node/2109.

Elastic strain might be because of chemical bond stretching, crystal structure deformation, or bond angle deformation. In the elastomer under a strain, the bond isn't elongated and the bond angles aren't deformed. Stretch of the elastomer is dependent upon rotation around the bonds that are changed to dihedral angles. The unstrained elastomer will prevail in the random coil structure. With the increase in strain, the molecules will unravel to the restraining linear structure. Thus, for a substance to be an elastomer, it must comprise the macromolecules. Large strain needs quite long molecules for uncoiling to be considerable. The creation of the unstrained random coil gives the meaning that the elastomer should be non-crystalline as any of the regular crystal structures won't be able to add to the elastomeric properties (Shanks, 2013). Figure 1.2 demonstrates the partly coiled 15 mer of poly(cis-1,4-isoprene). The model will be much longer if linear, whereas it will be problematic to observe if in an entirely coiled conformation (Promma et al., 2009).

The large rescindable strain should be quick, which means the limiting intermolecular forces should be minimal. The elastomers will have the least hydrogen bonding or the polar functional groups that add to the intermolecular forces. The steric prevention to uncoiling must be negligible so that the elastomers are improbable to have stiff intra-chain groups or heavy pendant groups. This is why a majority of the usual elastomers comprise hydrocarbon macromolecules with high molar mass. An elastomer will thus be the polymer exposed to all of the molecular intricacies (Haupt and Sedlan, 2001).

An elastomer is hypothetically an ideal polymer. Elastomers are ideal polymers to understand the function and structure of polymers. All of the other polymer characteristics are eccentricities on the structure of elastomer. A property not highlighted by the discussion is strain reversibility. Crosslinking is the difficulty that should be existent to offer reversibility (Bokobza and Kolodziej, 2006).

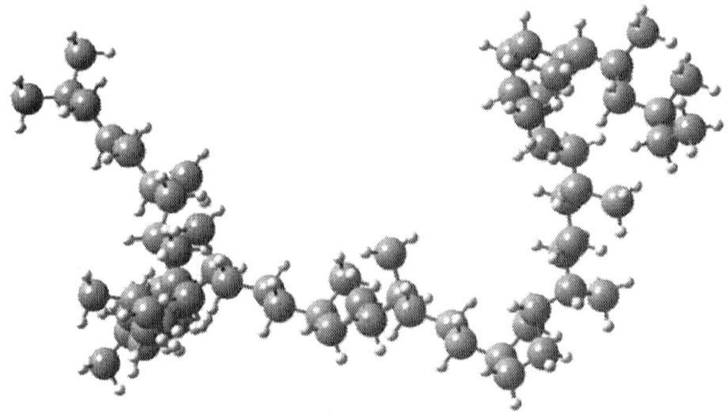

Figure 1.2. A poly(cis-1,4-isoprene) random coil having 15 monomer units.

Source: https://www.britannica.com/science/polyisoprene.

Figure 1.3 displays two molecules of poly(cis-1,4-isoprene) associated with crosslinks that prevent comparative translational motion, thus backing strain reversibility. This is very unlucky for studying elastomers since it means that the elastomers are insoluble, that the solution properties can't be measured (Wang et al., 2002; Rodas et al., 2014).

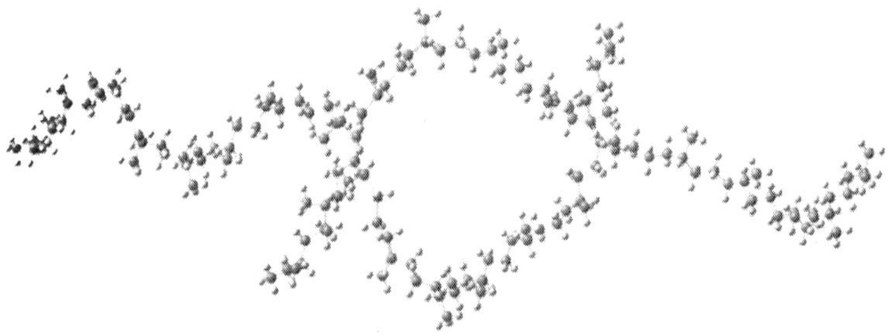

Figure 1.3. Two crosslinked 15 mers of a poly(cis-1,4-isoprene).

Source: https://www.springerprofessional.de/en/general-purpose-elastomers-structure-chemistry-physics-and-perfo/3636960.

Solution properties add much to the comprehension of macromolecules by implicit molar mass determination, hydrodynamic volume, solvent swelling, the mean distance between chain ends, theta solvent phenomenon, light scattering, and viscosity. Solution properties of the elastomers can normally be assessed before crosslinking to illustrate the behavior of all of the single molecules. Crosslinking can be performed to provide the minimum crosslinks needed for continuity all over the mass of elastomer that the crosslinks should at minimum give the percolation network all over. In the solvent, an elastomer will usually be crosslinked to or away from a gel point. Crosslinking to the percolation network will provide reversible strain. Crosslinking away from the percolation network will reduce elastomeric performance till it is absent in the highly crosslinked network. The solvent swelling is a suitable method to describe an elastomer as swelling utilizing solvation force is comparable to straining to utilize the physical force. Both straining and swelling are utilized to control the crosslink density as molar mass among crosslinks (Adam et al., 1989; Sare et al., 2001).

The behavior of elastomers, therefore far briefly presented, is why elastomers can be defined as ideal polymers. This chapter will describe the thermodynamics, polymerization, and structure techniques of these ideal molecules. The ideas will be extended to contain the less-than-ideal polymers that possess appropriate elastomeric properties for numerous applications, as those ideal polymers will dearth in extra properties needed of elastomers in particular or harsh environments. Instances with being presented that will comprise of formulation and uses of a product and advanced elastomers comprising blends and the interpenetrating networks (Wündrich, 1984; Ayoub et al., 2011).

1.2. THERMODYNAMICS

The use of thermodynamics to the elastomers necessitates that the molecules quickly equilibrate and conform with Hooke's law over the complete range of the strain being considered. Such kind of an elastomer is known as an ideal elastomer. The ideal elastomer instantaneously causes a collection of the ideal gas again. Thermodynamics of gases needed concern of nonconformity from ideal behavior. Gases diverge from ideality at high compression, because of the contact of atomic radii restraining compression and because of the non-zero intermolecular interactions (Zaharescu et al., 1998; Kumar and Sarangi, 2019). The elastomers deviate correspondingly at high extensions because of the molecular chain impending full-extension, crystallization of stretched molecular segments, and crosslinks becoming taut. At low extensions, preliminary unraveling of macromolecules is hindered by entanglements, some are permanent, whereas others can be unconstrained by unraveling of loops. At the high compressions, elastomers generally reach a boundary of compressibility because of impinging molecules or the entirely occupied volume (Sperling, 2005; Zhao et al., 2011).

Elastomer behavior ascends from the supportive segmental motions that are dependent upon the free volume into which the segments can normally move. The segmental molecular motions include cooperative pairs of the rotations of a bond; translational motion doesn't occur because of the crosslinks. Translational motion taking place with creep resembles the permanent set or a non-reversible constituent of the creep. Recoverable creep isn't elastomeric behavior as it is dependent on time, which is kinetic, whereas elastic behavior is reversible and immediate which is thermodynamic (Cantournet et al., 2009).

Free energy for the volume variation of a gas under continuous pressure is the Gibbs(G) free energy:

$$dG = dH - TdS;$$

whereas the work done on a system:

$$W = PdV + fdL \quad (1)$$

Free energy for the length variation of an elastomer under continuous volume (PDV = zero, supposing Poisson ratio = 0.5) is a Helmholtz (A) free energy:

$$dA = dU - TdS$$

whereas the work done on a system:

$$W = FdL \qquad (2)$$

where; U is the internal energy; H is the enthalpy; S is the entropy; V is the volume; T is the temperature; L is the length; F is the force; P is the pressure. The change of entropy with length is attained from Maxwell's equation (Ullman et al., 1993):

$$\left(\frac{dS}{dL}\right)_T = \left(\frac{-dF}{dT}\right)_L \qquad (3)$$

The temperature, length, and volume relationships for an elastomer and gas the disorder that is augmented by expansion as there occur more states that the molecules of gas can occupy, whereas with the compression of gas the number of states is restricted by lack of accessible free volume and as an outcome, the entropy is reduced (Chiang et al., 2012). The entropy of the elastomer is a disorder that is augmented with the contraction of an elastomer as the number of probable conformations of every macromolecule becomes infinity, whereas with the elongation of an elastomer the macromolecules usually become linear and the figure of conformations reaches one, an entirely linear macromolecule and also the entropy is least. Whereas the thermodynamics of gases triggers an engine compelled by volume and pressure change (PdV) and thermodynamics of the elastomers triggers an engine compelled by length and force change (FdL).

Thermodynamics of the gases envisages that as the gas is heated under continuous volume, the pressure of the gas will increase. An outcome of thermodynamics of the elastomers is the corresponding prediction that if the stretched elastomer is heated the length of elastomer will decrease. On the other hand, if the compressed gas when permitted to expand, the temperature of the gas will decrease. A similar is true for the stretched elastomer that is permitted to contract the temperature of the elastomer will decrease. If the gas is compressed, the temperature of a gas will increase, whereas if the elastomer is stretched, the temperature of the elastomer will increase (Figure 1.4) (Ullman, 1986; Elias et al., 2006).

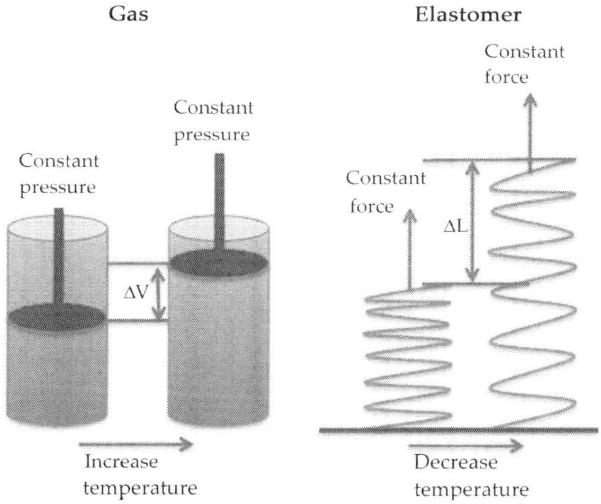

Figure 1.4. Temperature, length, and volume associations for an elastomer and gas.

Source: https://www.researchgate.net/figure/Temperature-volume-and-length-relationships-for-a-gas-and-elastomer_fig14_260751886.

Like most of the gasses, most elastomers aren't ideal. Non-ideal or the real elastomers don't behave entirely reversibly. Real elastomers exhibit dependence some time, where the elastomeric behavior is escorted by creep upon stretching and the viscoelastic recovery and everlasting set on contraction (Figure 1.5). The elastomers exhibit hysteresis upon relaxation or stretching. This supplementary real behavior yields hysteresis in stress versus strain curve; that is, the prolonged and contraction curves don't coincide. The area amongst the contraction and elongation curves is the energy that vanished during every cycle. The loss of energy is evident as heat (Chazeau et al., 2000; Li et al., 2014).

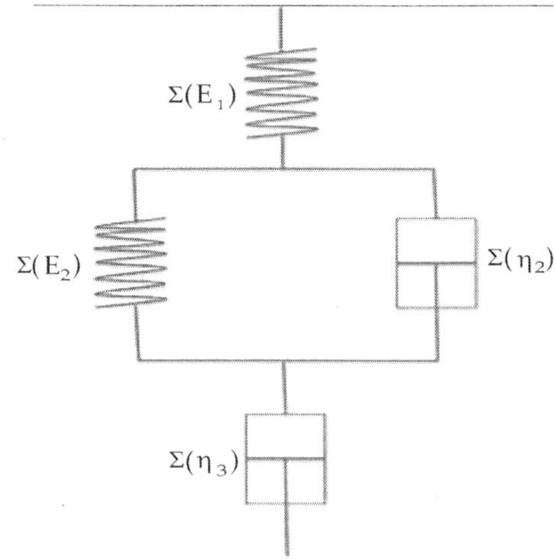

Figure 1.5. Viscoelastic, viscous, and elastic component models.

Source: https://www.springerprofessional.de/en/general-purpose-elastomers-structure-chemistry-physics-and-perfo/3636960.

Hysteresis is validated by the cyclic stress versus strain experiment. The hysteresis is the area amid the decreasing and increasing stress versus strain curves. In Figure 1.6, the solid lines demonstrate the low hysteresis of moderately crosslinked NR. The repetitive cyclic stress exhibits that the following stress-strain curves vary from the first stress-strain curve. This phenomenon is called the Mullins effect. Throughout the first straining, some of the entanglements can be eliminated therefore decreasing marginally the stress needed for following cycles. This can normally be designated as strain softening, in comparison to the strain hardening witnessed with the polyurethane (PU) elastomer. In Figure 1.6, the time response curves exhibit the reaction of NR to 6 recurrent cycles. The data are replotted as stress versus strain curves, and thus, the dotted curves exhibit the second and succeeding cycles. After the initial cycle, the response was normally reversible, and the dotted curves are accurately superimposed. If an NR is permitted to relax an analogous diverse first cycle can be observed again. This phenomenon is called strain conditioning (Brown et al., 1980; Liu et al., 2011). A series of almost six repetitive stress versus strain cycles was forced upon SBR and the outcomes are exhibited in Figure 1.7. The time-centered chart demonstrates the total data on stress and strain. The stress versus strain chart demonstrates the higher hysteresis of SBR when matched with NR, particularly for the first cycle. The curves of stress versus strain after the initial cycle were nearly precisely superimposed though not exact and with the greater hysteresis as compared to the outcomes for NR (Gamlin et al., 2000).

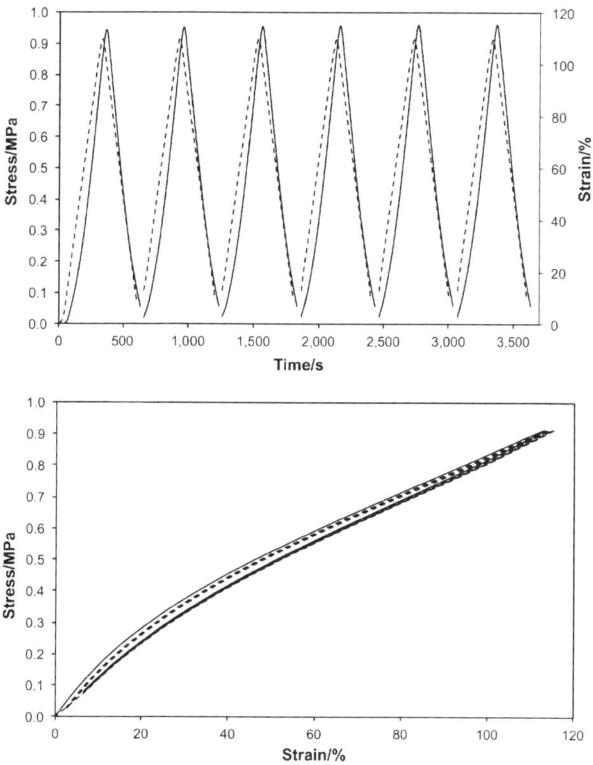

Figure 1.6. Stress versus strain time (upper) and the hysteresis (lower) curves for a poly(cis-1,4-isoprene) elastomer.

Source: https://www.researchgate.net/figure/Stress-strain-time-upper-and-hysteresis-lower-curves-for-polycis-1-4-isoprene_fig8_260751886.

The variation of the initial cycle to the following cycles, the Mullins effect, was quite greater with SBR as compared to NR. SBR is more wear-resistant and harder than the NR, with a substantially improved hysteresis. The energy vanished by hysteresis during the cyclic stress is disappeared as heat so that the high hysteresis elastomers generally have a detrimental heat generated. This is a benefit when the stress damping is needed, but not when the high heat generated will cause a collapse of the elastomer. The damping behavior of elastomers can usually be improved by the inclusion of the fillers. Usual filters are silica, talc, and carbon black (CB) (Mooney, 1948).

The elastomer hysteresis is where the kinetics imposes upon thermodynamics. The kinetics is a response that is time dependent, which with the macromolecules is temperature-dependent.

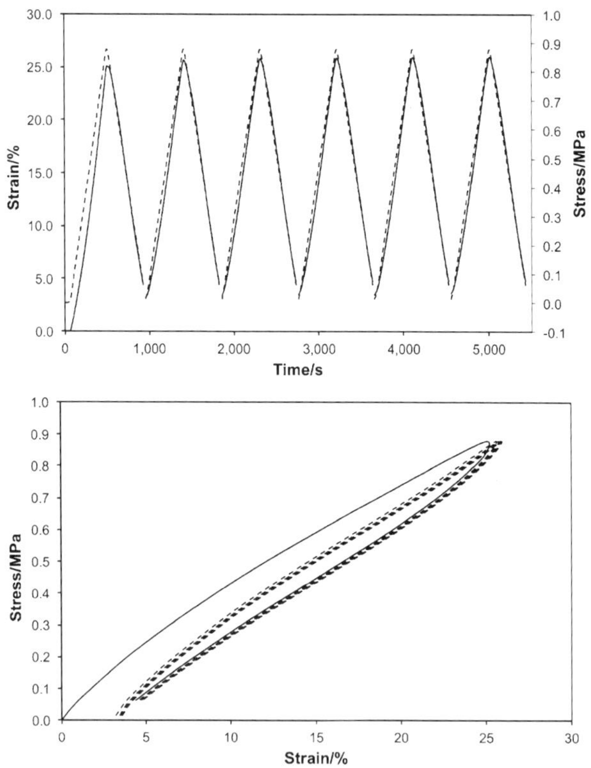

Figure 1.7. Stress versus strain time (upper) and the hysteresis (lower) curves for the poly(styrene-co-butadiene) elastomer.

Source: http://springer.nl.go.kr/chapter/10.1007%2F978-3-642-20925-3_2.

1.3. KINETICS

Elastomer kinetics is an outcome of the non-ideal behavior that is exhibited in the former section outcomes in hysteresis and production of heat. Another result of the kinetics is the transition of glass or amorphous polymers or the amorphous regions inside the semi-crystalline polymers. The transition of glass takes place over the temperature range stated as Tg (glass transition temperature). When a polymer's temperature is below the glass transition temperature, there is inadequate thermal energy to overwhelm the energy of activation for the segmental motions and a polymer will normally be glassy (Van Amerongen, 1950; Cella, 1973). When the polymer's temperature is above the glass transition temperature, segmental motions will take place and a polymer will show elastomer behavior. When the polymer's temperature is in the range of glass transition, then the polymer will be dominantly viscoelastic triggering damping of forces with substantial conversion to heat. In order to be an efficient elastomer, the polymer must be quite above its glass transition temperature to diminish viscoelastic behavior. The elastomeric performance at surrounding temperatures (20 to 25 C) efficiently needs glass transition temperature in the range of –70 to –20°C for an efficient immediate reversible response to take place (Hong et al., 2010; Tikhomirov et al., 2018).

Creep and the recovery tests were carried out on SBR and NR to illustrate that creep is nearly non-existent in the crosslinked elastomers and that the recovery is quick and almost complete. After the first immediate elastic strain, the creep of the NR is low because of the network of crosslinking, whereas recovery was immediate to the strain of 1%, followed by nearly complete recovery with the low permanent set (Figure 1.8). A comparative analysis of the SBR creeps displayed a somewhat slower elastic response trailed by a greater viscoelastic creep. The recovery of SBR was quick, although with a permanent set (Figure 1.9).

Measurement of the stress versus strain response after the large pre-strain has discoursed under the curves of hysteresis of Figures 1.6 and 1.7, whereas another technique is to measure the modified stress response of the elastomer on the highly strained elastomer. Such kind of analysis displayed that the modulus of loss was free of a fundamental shear to strain over the range of engineering strain for unfilled and the CB-filled compounds of rubber. These outcomes are beneficial for understanding the dissipation of energy by strained elastomers (Suphadon et al., 2010; Lin et al., 2019).

1.4. STRUCTURE

The molecular structure of the elastomer comprises random coils linked by crosslinks. Unevenness is essential to avert crystallinity. Unevenness can be because of the geometric isomers where the configuration of cis gives ideal elastomer properties, although the double bond confines motion of the carbon atoms bonded. The trans configuration adds to the regular planar zig-zag compliance that is crystallizable. An entirely saturated hydrocarbon polymer might be elastomeric if the alternatives are in the atactic configuration, or if the polymer is a casual copolymer where the segments can't co-crystallize (LeBaron and Pinnavaia, 2001; Wang et al., 2012). The statistical explanation of the polymer elasticity is centered upon the casual circulation of chain links creating the random coil conformation of macromolecules. A casual coil is more stable thermodynamically as compared with the completely extended chain since there exists an infinite number of random coils, whereas there is just one completely extended chain. The casual coil figures are self-averting in that the model contemplates the omitted volume of a molecule. The supposition of the random coil conformation permits extrapolation of structural features including radius of gyration, end-to-end distance, contour length, characteristic ratio, and persistence length (Elias et al., 1990).

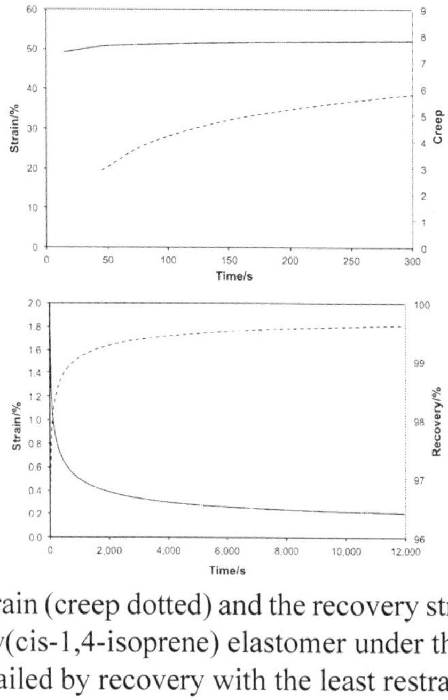

Figure 1.8. Creep strain (creep dotted) and the recovery strain (recovery dotted) response for the poly(cis-1,4-isoprene) elastomer under the ductile creep stress of nearly 0.5 MPa trailed by recovery with the least restraining force.

Source: https://www.springer.com/gp/book/9783642209246.

Elastomer deformational reversibility needs molecular crosslinking away from gelation so that comparative molecular positions are static during conformational variations. Gelation takes place when the crosslinked molecules spread all over the system creating the percolation network (Kim et al., 2003; Myung et al., 2008). The crosslinking of covalent bonds will avert comparative translational motions. The ionic crosslinks can normally be introduced utilizing covalent metal ions with the polymer having anionic charges alongside the chain. On the other hand, physical crosslinks will usually be appropriate for reversibility. The physical crosslinks can usually be because of the phase alienated block copolymer structure where the separate phase will crystallize or glassy (high Tg).

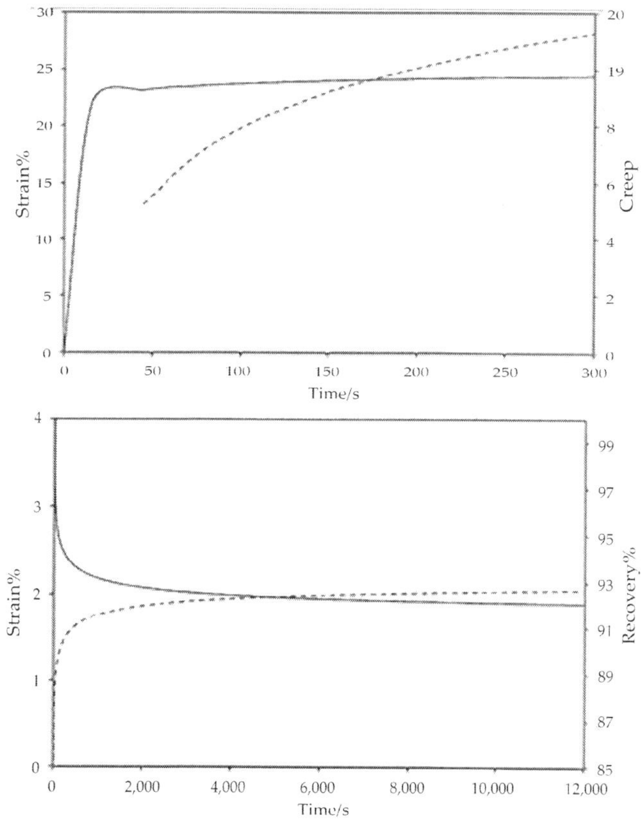

Figure 1.9. Creep strain (creep dotted) and the recovery strain (recovery dotted) response for the poly(butadiene-co-styrene) elastomer under the ductile creep stress of almost 1.0 MPa trailed by the least restraining force.

Source: https://www.springer.com/gp/book/9783642209246.

1.5. POLYMERIZATION

Elastomers are primarily synthesized utilizing chain-growth polymerization since high molar mass is the characteristic of this procedure and the high molar mass is needed for high chain addition. Elastomers synthesized with the help of procedure having radical initiation are the hydrocarbon kinds like polybutadiene, poly(butadiene-co-acrylonitrile), fluorocarbon, and poly (butadiene-costyrene) elastomers. Polyisobutylene is started utilizing cations. The block copolymers poly(styrene-butadiene-b-styrene) is started by anions. The poly(ethylene-co-propylene) and the diene terpolymers are synthesized by initiation with the co-ordination metal catalysts (Mazurek et al., 2014; Akram et al., 2019).

The elastomers synthesized with the help of step-growth procedures include polysulfides, polysiloxanes, and the silicone or PUs elastomers. Every key chemical structure kind of elastomer is discussed in the following section (Datta, 2004).

1.6. PARTICULAR ELASTOMER STRUCTURE AND ITS PROPERTIES

1.6.1. Aromatic and Aliphatic Hydrocarbon Elastomers

- **Natural Rubber (NR):** It is an elastomer having the fundamental monomer of cis-1,4-isoprene. NR is developed by treating the fluid of rubber tree with steam and then compounding it with the vulcanizing agents, fillers, and antioxidants. NR is broadly utilized for applications needing wear resistance or abrasion, damping, and electric resistance, or shock-absorbing properties like large truck tires, aircraft tires, and off-the-road giant tires. It is resistant chemically to acids, alcohol, and alkalis. However, it doesn't perform well with the oxidizing chemicals, ozone, oils, atmospheric O_2, petroleum, ketones, and benzene (Sivakumar and Gopal, 2011; Mahaling et al., 2015).
- **Styrene-Butadiene Rubber (SBR):** It is the synthetic rubber copolymer comprising of butadiene and styrene. The chemical resistance of SBR is comparable to that of NR, though, it displays an outstanding abrasion resistance than NR and polybutadiene that makes SBR an appropriate material for automobile tires (Logothetis, 1989; Rogulska and Kultys, 2016).

- **Butyl Rubber (IIR):** It is the copolymer of isoprene and isobutylene as fundamental monomer units. This kind of synthetic rubber possesses a very low rate of permeability, making it a considerable seal under the vacuum. It also possesses good shock dampening and electrical properties. It is utilized in various applications needing an airtight rubber. The main applications of IIR are tire inner hoses and tubes (Masamoto and Kubo, 1996; Wang et al., 2020).
- **Ethylene Propylene Diene Monomer (EPDM):** It is a synthetic rubber comprising of propylene and ethylene. It has excellent heat, weather, and ozone resistance because of its stability and the saturated polymer support structure. As the non-polar elastomers, EPDM possesses good electrical resistivity, along with resistance to the polar solvents, like water, alkalis, acids, phosphate esters, and several alcohols and ketones (Figure 1.10). It is primarily utilized as the typical lining material for a radiator, automotive weather-stripping and seals, electrical insulation, roofing membrane, and steam hoses. Properties of several hydrocarbon elastomers are exhibited in Table 1.1 (Jeong et al., 2016).

1.6.2. Nitrile and Halogen Replaced Elastomers

CR (Polychloroprene) was among the first commercially effective synthetic rubbers having a yearly consumption of around 3,00,000 tons around the globe (excluding PR of China and the former Soviet Union). It is the chlorinated rubber material, which was made in the year 1932 by Collins, Carothers, and co-workers utilizing emulsion polymerization methods. In a similar year, DuPont started advertising the polymer under the name Duprene and from 1938 under the name Neoprene. From the start till the 1960s, chloroprene was made by the acetylene process, which needed costly feedstock. The current chloroprene process is economical and quite safer. Chloroprene is made in 3 steps from readily accessible butadiene (Tang and Gupta, 1988):

1. Chlorination;
2. Isomerization; and
3. Dehydrochlorination.

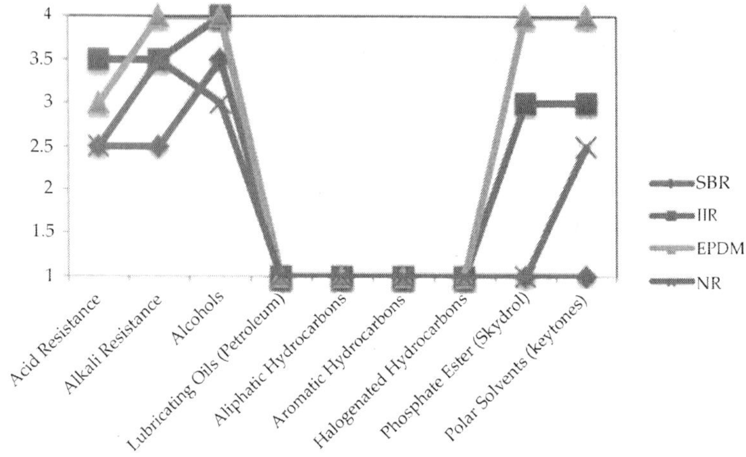

Figure 1.10. A graphic demonstration of the hydrocarbon elastomer properties (4 excellent; 3 good; 2 fair; 1 poor).

Source: https://www.springer.com/gp/book/9783642209246.

Table 1.1. Comparison of the Key Hydrocarbon Elastomers (4 Excellent; 3 Good; 2 Fair; 1 Poor)

Elastomers	***EPDM***	***IIR***	***SBR***	***NR***
Economy (cost)	*1*	*3*	*1*	*2*
Resilience/rebound	*2*	*4*	*2*	*1*
Tensile strength	*1.5*	*2*	*1.5*	*1*
Adhesion to metals	*2.5*	*2*	*1*	*1*
Compression set	*1.5*	*2.5*	*1.5*	*2*
Adhesion to metals	*2.5*	*2*	*1*	*1*
Abrasion resistance	*1.5*	*2.5*	*1.5*	*1*
Tear resistance	*1.5*	*2*	*2*	*1.5*
Dynamic properties	*1.5*	*3*	*2.5*	*1*
Ozone resistance	*1*	*1.5*	*4*	*4*
Water swell resistance	*1*	*1.5*	*1.5*	*1*
Electrical properties	*2*	*2*	*2*	*2*
Flame resistance	*4*	*4*	*4*	*4*
Steam resistance	*1*	*2*	*4*	*4*
Gas impermeability	*2*	*1*	*2.5*	*3*

Originally made as to the oil-resistant alternative for NR, Polychloroprene has quite good resistance towards several organic chemicals including gasoline, mineral oils, and some halogenated or aromatic solvents. It has excellent aging resistance, high weather, and ozone resistance. In comparison to most of the other types of rubber, polychloroprene exhibits an astonishingly higher level of confrontation to microorganisms, like bacteria and fungi. Furthermore, it has very low flammability and exceptional resistance to the damage caused by twisting and flexing, a higher toughness (Figure 1.11). Thus, its amalgamation of exclusive properties makes it suitable for numerous applications all over the industry. Polychloroprene is normally utilized as hose covers. In the application of construction, it has been utilized for several years as the selected elastomer for bearings in bridges and machinery, seals, and bellows. The compound is aimed to have outstanding low temperature, ozone, weather, and flex resistance. In the automotive industry, it aids as gaskets, boots, seals, power transmission belts, and air springs, formed, and squeezed goods, and cellular products sealants and adhesives. Some of the other uses are insulating sockets of CPU, to develop waterproof automotive covers of the seat, rollers for the textile and printing industry (Ula et al., 2018).

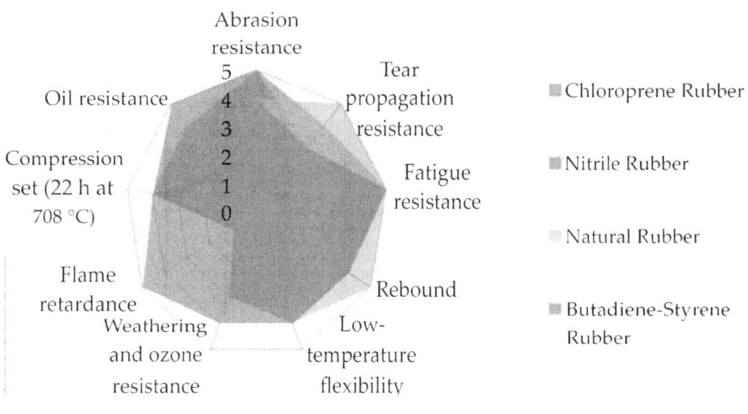

Figure 1.11. Property contrast of chloroprene and nitrile elastomers (5 excellent; 4 very good; 3 good; 2 fair; 1 unsatisfactory). The ratings are dependent on compound composition; hence all of the optimal values might not be attained at the same time).

Source: https://www.routledge.com/Handbook-of-Specialty-Elastomers/Klingender/p/book/9780367387808.

Acrylonitrile butadiene rubber (NBR), also called nitrile rubber, is the synthetic rubber made from the copolymer of acrylonitrile and butadiene. It was developed by the chemists Erich Konrad, Eduard Tschunkur, and Helmut Kleiner, and the patent for this novel oil-revolting rubber was given on April 26, 1930 (Steffen et al., 2018; Cheng et al., 2019).

The chemical and physical properties of the nitrile rubber change depending on the conformation of acrylonitrile, which usually ranges from 15–50%. With increasing content of acrylonitrile, the rubber exhibits greater confrontation to swelling by the hydrocarbon oils, lower permeability to the gases, and higher strength. Though, the higher Tg makes the rubber less stretchy at the lower temperatures (Petrović and Ferguson, 1991).

The most usual nitrile rubber having acrylonitrile content of 31 to 35% withstands the temperature range of nearly –40–107°C. Nitrile rubber is well-thought-out to be the main oil, heat, and fuel resistant elastomer around the globe. It is also resilient towards aliphatic hydrocarbons. Nevertheless, it isn't resilient towards aromatic hydrocarbon, ester, ketones, and strong oxidizing chemicals. Nitrile rubber has very poor resistance to sunlight, weathering, and ozone. Nitrile rubber is broadly utilized in the automotive industry as the automotive gaskets and seals, which subject to interaction with the hot oils. It is utilized as the automotive water treating applications and in oil and fuel handling hose too. In the healthcare industry, its pliability makes it the ideal material for non-latex gloves. Other uses of the nitrile rubber comprise the rolls for scattering ink in printing and the hoses for the oil products, as a pigment binder and an adhesive (Blackwell and Gardner, 1979).

1.6.3. Sulfide Elastomers

Polysulfide (PSR) is the group of chemical compounds comprising a chain of sulfur atoms. PSR was first discovered and patented by Nathan Mnookin and Joseph C. Patrick by coincidence when they were struggling to discover cheap antifreeze. They called it Thiokol and the production started in 1929 (Krol, 2007).

Polysulfide displays outstanding chemical resistance towards greases and oils and they have good dielectric properties. Some of the other unique properties of polysulfide including dimensional stability, low moisture vapor transmission, flexibility, weather ability, and low gas transmission. These properties make PSR particularly beneficial in a range of sealant uses (Gunatillake et al., 2003).

The present kinds of PSR are Thiokol ST, Thiokol LP, and Thiokol FA. Thiokol FA is the tradename for a polysulfide copolymer created from ethylene dichloride and di-2-chloroethyl formal. With its outstanding solvent resistance, Thiokol FA is suitable for roller applications needing resistance to highly aromatic solvents, ketones, and some of the chlorinated solvents. It is an ideal choice of hose tubes and paint spray can gaskets too (Guenthner and Stivers, 1984). Thiokol ST is a branched polysulfide created from di-2-chloroethyl formal having about 2% 1,2,3-trichloropropane as the trifunctional branching units, aimed for mechanical goods. The contrast of properties of Thiokol ST and Thiokol FA is displayed in Table 1.2. Thiokol LP, a name for a variety of the liquid polysulfide rubbers (PSRs) attained by cleavage of Thiokol ST and Thiokol FA, which is among the most broadly utilized mercaptan-terminated polymers. Thiokol LP has been utilized as a base polymer in sealants since the 1950s. With nearly more than 50 years of experience, liquid polysulfide has been proven effective. The excellent resistance of polysulfide to petroleum products has made PST the typical sealant for nearly all of the aircraft's integral bodies and fuel tanks. It is also utilized in insulating construction sealants and glass window sealants (MacLachlan, 1978).

Table 1.2. The Contrast of the Mechanical Properties of 2 Polysulfide Elastomers

Physical Properties	**Thiokol FA**	**Thiokol ST**
Shore (A)	65–70	65–70
Elongation (%)	380	220
Modulus at 100% (MPa)	5.1	3.7
Gehman low temperature, 8°C	–45	–50
Compression set 22 h at 70°C	100	20
Tensile strength (MPa)	8.3	8.3

Source: Klingender (2008).

1.6.4. Polyurethane Elastomers

PUs are the most famous polymers utilized to make foams. For preparing PU, the raw materials are polyisocyanates, diamines, catalysts, polyols, blocking agents, and additives. Usually, they are made by the reaction of diol and diisocyanate with either ether or ester backbone in the existence of catalysts. Ether-centered PUs are utilized to produce rigid and flexible foams, whereas ester-centered PUs are utilized to produce flexible foams, coatings, and elastomers (Borgmann et al., 2011).

The structure of PU is highly inclined by the intermolecular forces, like hydrogen bonding, van der Waals forces, polarizability, crosslinking, and stiffness of the chain. There might be crystalline regions amongst the flexible chains. The polymers display low corrosion resistance to the strong alkalis and acids, and the organic solvent. The flexible foams are utilized for several domestic applications, like sofas, carpet backs, car seats, and cushions, whereas rigid foams are utilized as the materials of thermal insulation for transportation of frozen food products and cryogenic fluid. Some other uses of PUs are shoe soles in the shoe company, bumper covers, dashboards, fenders, and moldings in the automobile industry (Gonzaga et al., 2009).

The cyclic stress versus strain response for the polyether-urethane is demonstrated in Figure 1.12. The time chart for data displays the regular variations of strain with the stress of 6 cycles. The stress versus strain chart demonstrates high hysteresis displayed by the PU, because of its polar structure giving intermolecular interactions that trigger deviation from the perfect elastomer response. The 2nd and succeeding cycles vary from the initial cycle the similar as observed for SBR and NR except that the strain hardening has taken place with the PU and that the recurring cycles aren't so precisely overlayed. With the increase in strain and elongation of the random coils, intermolecular interactions become favorable, and therefore the stress needed for any specific strain has augmented. There has been the memory of increased collaborations after the initial cycle, even though they remain persistent for following cycles (Courval and Gray, 1975; Sotayo et al., 2020).

The cyclic stress response of the PU series has been determined into inelastic and elastic contributions. The initial stress cycle dissimilarity was increased with decreased soft and hard phase separation. In the following cycles, the Mullins effect was accredited to the separation of the chain segments from the isolated hard phase to upsurge coupling with the soft constant phase (Buckley et al., 2010).

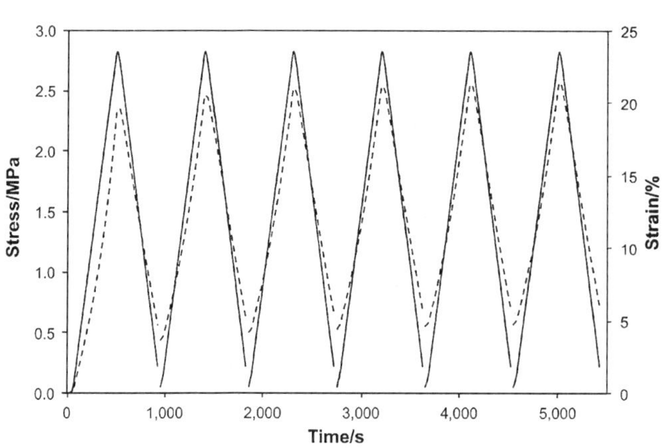

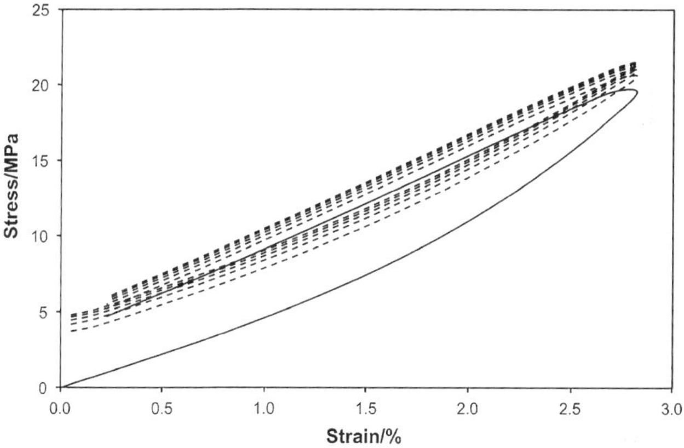

Figure 1.12. Stress versus strain time and the hysteresis curves for the poly(ether-co-urethane) thermoplastic elastomer.

Source: https://www.researchgate.net/figure/2-Stress-strain-time-upper-and-hysteresis-lower-curves-for-a-polyether-co-urethane_fig2_260751886.

The thermoplastic PU elastomers are centered on either polyester or polyether polyol pre-polymers. Polyesters are usually derived from adipate or succinate monomers pooled with ethylene, butane, propylene, or dimers thereof centered diols. The succinates can be obtained from sugars forming interest due to their bio-origin. The adipate and succinate polyester-polyurethane were matched and discovered to be analogous, with somewhat higher Tg, increased interaction amongst soft and hard phases, and to some extent lower abrasion resistance (Sonnenschein et al., 2010) (Tables 1.3 and 1.4).

Table 1.3. Structure and Abbreviation of Elastomers

Common Name	Abbreviation	Structure of Repeat Unit
cis-1,4-polyisoprene	NR, caou-tchouc	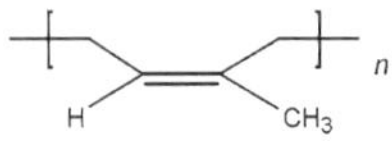
1,2-polybutadiene	BR	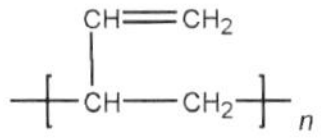
trans-1,4-polyisoprene	Gutta-percha	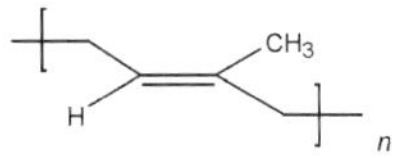
Butyl rubber	IIR	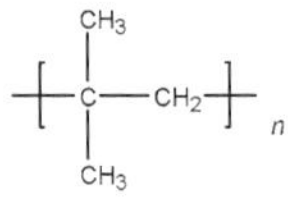
cis-1,4-polybutadiene trans-1,4-polybutadiene Butadiene styrene or styrene-butadiene rubber	BS/SBR	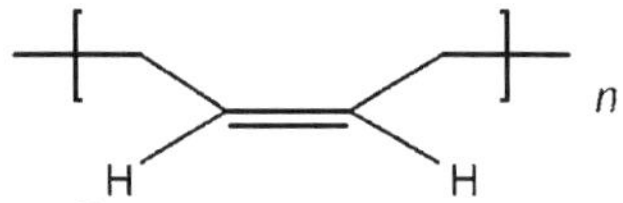 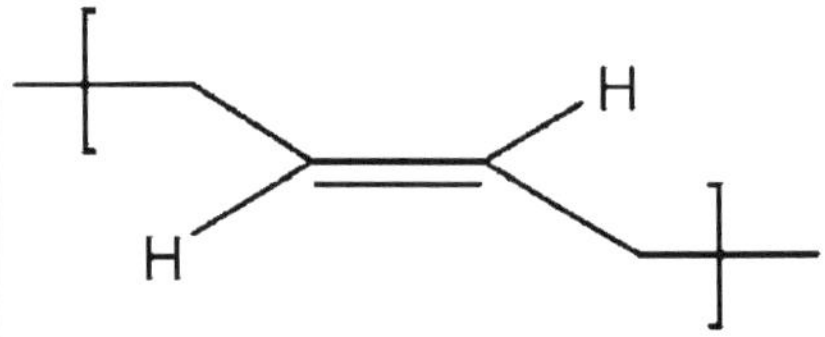 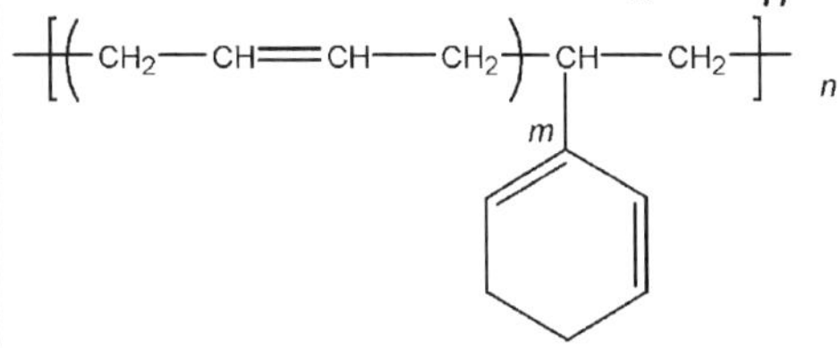
Epichlorohydrin rubber	ECO	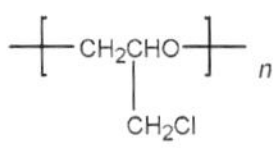
cis-1,4-polychloroprene	CR	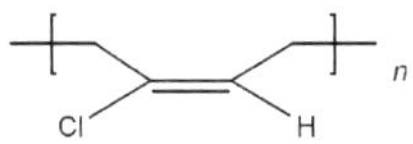
trans-1,4-polychloro-prene Polysulfide butadiene rubber	PSR	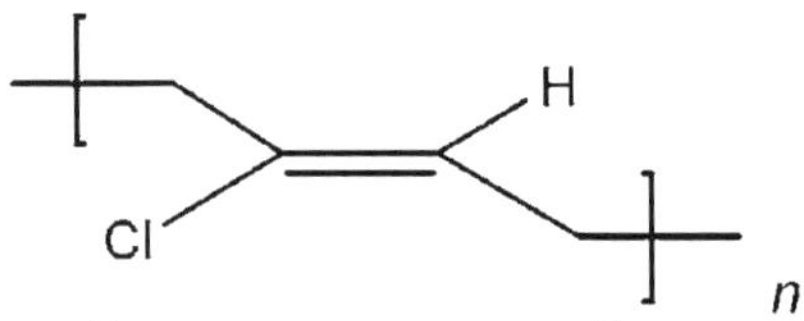 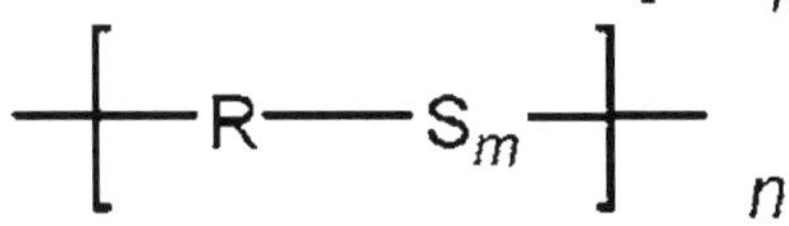

Acrylonitrile butadiene rubber	NBR	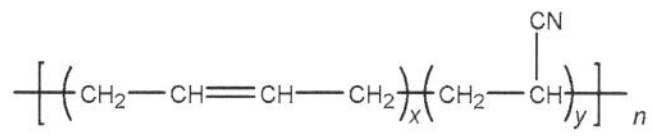
Ethylene propylene diene rubbers	EPDM	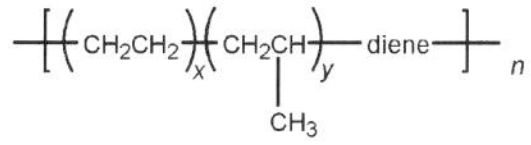
Silicone rubber	SR	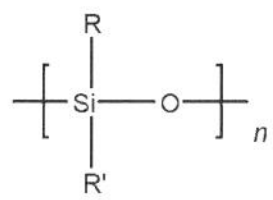
Polyurethanes	PU	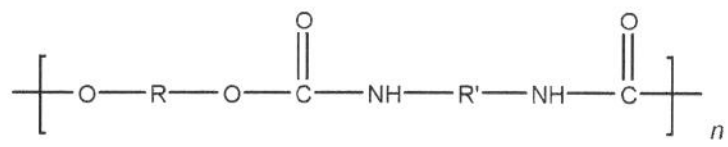

Note: R' and R: aryl or alkyl.

Table 1.4. Chemical and Mechanical Properties of Elastomers

Common Names	Density (ρ/kg m^{-3})	Elastic or Young's Modulus (E/GPa)	Ultimate Tensile Strength (σ_{UTS}/MPa)	Yield Tensile Strength (σ_{ys}/MPa)	Elongation at Break (Z/%)	Shore A or D Hardness	Minimum Operating Temperature Range	Maximum Operating Temperature Range	Glass Transition Temperature (T_g/°C)	Specific Heat Capacity (c_p/J Kg^{-1} K^{-1})	Thermal Conductivity (k/W m^{-1}K^{-1})	Refractive Index (n_D/nil)	Relative Electric Permittivity (@1 MHz) ($\varepsilon\rho$/nil)	Electrical Resistivity (ρ/ohm cm)
cis-1,4-polyisoprene	920–1037	3.3–5.9	29	17.1–31.7	660–850	A30-95	–56	82	–73	1,830	0.15	1.519–1.52	n.a.	2.6
trans-1,4-polyisoprene	n.a.	n.a.	n.a.	n.a.	n.a.	n.a.	n.a.	n.a.	–58	n.a.	n.a.	1.509	n.a.	n.a.
1,2-polybutadiene	910	2.1–10.3	n.a.	13.8–17.2	450	A45–80	–100	95	n.a.	n.a.	n.a.	1.5	2.5	1,E + 15
cis-1,4-polybutadiene	n.a.	n.a.	n.a.	n.a.	n.a.	n.a.	n.a.	–102	n.a.	1,854	n.a.	1.52	n.a.	n.a.
trans-1,4-polybutadiene	n.a.	n.a.	n.a.	n.a.	n.a.	n.a.	n.a.	–58	n.a.	2,402	n.a.	1.518	n.a.	n.a.
Butadiene styrene or styrene-butadiene rubber	940	2.1–10.3	21	12.4–20.7	450–500	A30–90	–60	120	n.a.	n.a.	n.a.	1.53	2.4	1,E + 14
Butyl rubber	917	0.3–3.4	17	n.a.	700–950	A30–100	–45	150	–75 to 67	1,950	0.13–0.23	1.5081	n.a.	n.a.
cis-1,4-polychloroprene	1230–1250	07–20.1	n.a.	3.4–24.1	100–800	A30–95	–43	107	–20	2,170	0.192	1.554–1.558	2.0–6.3	1,E + 11
trans-1,4-polychloroprene	n.a.	n.a.	n.a.	n.a.	n.a.	n.a.	n.a.	–40	n.a.	n.a.	n.a.	n.a.	n.a.	n.a.

Polysulfide butadiene rubber	1340	n.a.	4.83–8.63	n.a.	100–400	A65–70	–54	100–400	–55	n.a.	n.a.	1.6–1.7	1.3	1,E + 08
Acrylonitrile butadiene rubber	1000	n.a.	21	n.a.	510	A30–90	–40	121	n.a.	n.a.	n.a.	1.52	2.5	1,E + 15
Ethylene propylene diene rubber	850	n.a.	21	n.a.	100–300	A30–90	–51	150	n.a.	n.a.	2.22	1.474	2.5	n.a.
Epichlorohydrin rubber	1270	n.a.	17	n.a.	400	A60–90	–46	121	n.a.	n.a.	n.a.	n.a.	n.a.	n.a.
Polyurethanes	1050–1250	n.a.	29–49	n.a.	10–21	n.a.	n.a.	n.a.	n.a.	1,800	0.21	n.a.	n.a.	1,E + 12
Silicone rubber	n.a.	n.a.	65	n.a.	n.a.	D60A	n.a.	n.a.	n.a.	n.a.	n.a.	n.a.	n.a.	n.a.

1.6.5. Fluorocarbon Elastomers

Fluorocarbon elastomers (FKM) are synthetic copolymers obtained by substituting some or all hydrogen atoms in the hydrocarbons with Fl (fluorine) atoms. They are usually utilized when the other elastomers flop in unfriendly environments. The two key significant properties of this group of elastomers are heat resistance and chemical resistance. The unusual heat and chemical resistance of FKM can be accredited to the high energy of bond of C-F bond and also the higher bond energy of C-C because of the existence of electronegative fluorine. The chemical resistance of FKM versus the other elastomers is demonstrated in the radar chart in Figure 1.13.

FKM is utilized broadly in the severe environment in the aerospace and automotive industry (Figure 1.14). In the automotive industry, FKM has been developed into several items, like shaft seals, O-rings, valve stem seals, engine head gaskets, diaphragms for fuel pumps, filter casing gaskets, water pump gaskets, seals for the exhaust gas and contamination control equipment, fuel hoses, lubricating circuits, bellows for turbo-charger, etc. Military and aerospace application are for hydraulic hoses, shaft seals, O-rings, gaskets for firewalls, electrical connectors, fuel tanks, traps for the hot engine lubricants, heat-shrinkable tubing for insulation of the wire, etc., petrochemical, and chemical plants use O-rings, diaphragms, expansion joints, blow-out preventers, gaskets, valve seats, hoses, safety gloves and clothes, duct coatings and stack, and lining of the tank (Shanks, 2013).

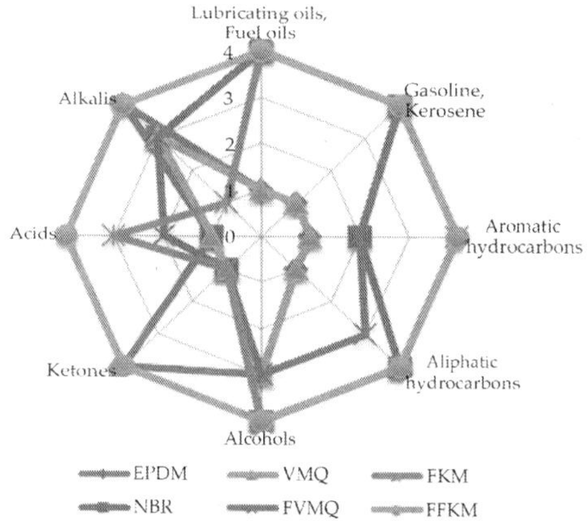

Figure 1.13. Chemical resistance of the elastomers compared with fluorocarbon elastomers.

Source: https://www.amazon.com/Handbook-Specialty-Elastomers-Robert-Klingender/dp/1574446762.

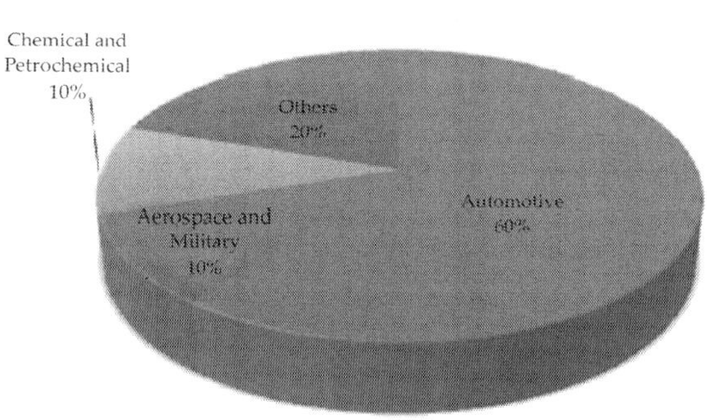

Figure 1.14. Relative market segments of the fluorocarbon elastomers displaying specialty automotive supremacy.

Source: https://www.wiley.com/en-us/Modern+Fluoropolymers%3A+High+Performance+Polymers+for+Diverse+Applications-p-9780471970552.

1.6.6. Polysiloxane Elastomers

These are the mixed organic-inorganic polymers, which are quite famous under the name of silicone rubber (SR). Rather than the skeleton of the organic carbon chain, these materials are centered on silicon and O_2 support. Berzelius found silicon, the foundation of silicones in the year 1824, from a reduction of the silicon tetrafluoride with potassium (K). In the 1940s, F.S. Kipping accomplished a wide-ranging synthesis of the silicone compounds and invented the name 'silicones.'

Polydimethylsiloxane (PDMS) is the most common siloxane. It can be made from the hydrolysis of the dimethyldichlorosilane that produces a silanol. The subsequent silanol instantaneously experiences polycondensation. The procedure of the hydrolysis and composition of the silane mixture govern the size of polysiloxane molecules. Usually, silicones have outstanding resistance to cold and heat, varying from –60–+250°C. They exhibit hydrophobic properties which empower them to behave as release agents for the tacky substances. Silicones are resistant to radiation and ozone. The low elasticity modulus and the high mechanical degeneracy factor make these materials an outstanding medium for sound and shock absorption. In the field of cosmetics, silicones are broadly utilized in various skin-care products, shaving products, personal lubricants, and hair products due to their compatibility with skin, pleasant touch, and water-repellent effect. In the automotive industry, silicones are utilized as the lubricant for the components of a brake, gaskets in the automotive engines, and also for airbags as sealants and coatings. Transparent tubes treated from silicones are utilized for the medical applications and food industry and the insulation of cable for usage at the high temperature (Polmanteer and Hunter, 1959).

1.7. THERMOPLASTIC ELASTOMER (TPE) STRUCTURE AND PROPERTIES

These are exceptional synthetic compounds that syndicate some properties of the rubber with the treating benefits of thermoplastics. Normally, they can be characterized into two groups: blends and multi-block copolymers. The first group comprises hard thermoplastic blocks and soft elastomers, like styrene block copolymers (SBCs), polyamide-elastomer block copolymers (COPAs), polyether ester-elastomer block copolymers (COPEs), and polyurethane—elastomer block copolymers (TPUs). Thermoplastic blends can usually be distributed into polyolefin blends (TPOs) and TPVs (dynamically vulcanized blends). The world TPEs request by kinds of TPE and by the regions in the years 2014, 2009, and 2004 are displayed in Figure 1.15 (Breiner et al., 1999).

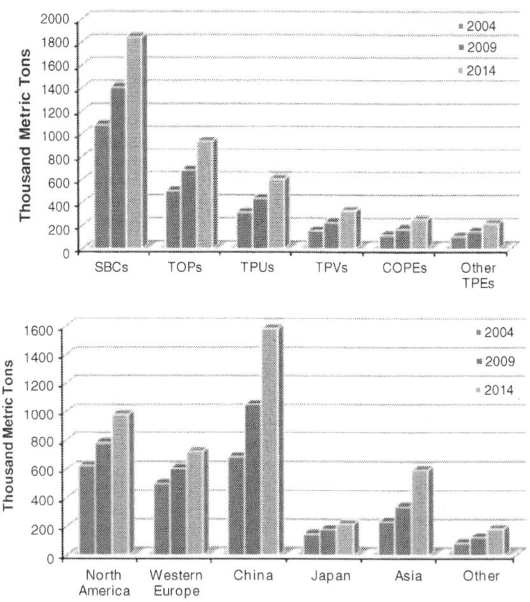

Figure 1.15. World TPE demand in 2014, 2009, and 2004 by types of thermoplastic elastomers and by regions.

Source: https://www.sciencedirect.com/book/9780815515494/handbook-of-thermoplastic-elastomers.

1.7.1. Styrenic Block Copolymer

SBCs (Styrenic block copolymers) are the largest with respect to volume and the cheapest group of TPEs. SBCs are centered on the simple molecules (A-B-A kind) that comprise three blocks at minimum, specifically one soft elastomeric midblock and two hard polystyrene end blocks. The midblock is normally a polydiene, either polyisoprene or polybutadiene, occasioning in a famous family of styrene-butadiene-styrene (SBS) and styrene-isoprene-styrene (SIS). Other styrenic block copolymers that have been successful commercially include SEPS (ethylene—propylene), SIBS (polyisobutylene), SEEPS (ethylene—ethylene-propylene), and SEBS (ethylene—butylene). The structure of styrenic block copolymers is schematically shown in Figure 1.16 (Wang et al., 2016; Abdeen, 2017).

The tensile strength of styrenic block copolymers is much higher as compared to those measured on the unreinforced vulcanized rubbers. The majority of the styrenic block copolymers have an elongation at breakdown ranges over 800% and resilience is analogous to that of the vulcanized rubbers. Styrenic block copolymers display behavior of nonnewtonian flow due to their severe segmental incompatibility. SBCs have melt viscosity much higher as compared to those of polyisoprene, polybutadienes, and random copolymers of butadiene and styrene. Styrenic block copolymers are seldom utilized as pure materials. The majority of the styrenic block copolymers can be mixed readily with the other polymers, fillers, oil, colorants, resins, and processing supports to meet the desired mechanical and physical properties (Polacco et al., 2006; Wang et al., 2016).

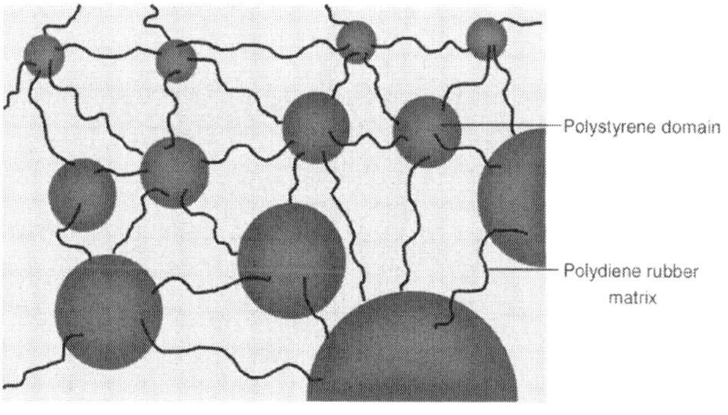

Figure 1.16. Diagram of the (styrene-diene-styrene) block copolymer.

Source: https://www.growkudos.com/publications/10.1007%2525 2F978-3-642-20925-3_2/reader.

The main applications for styrenic block copolymers are footwear and sealants and adhesives. They are also utilized in adjusting the asphalt's performance for roads and roofing, mainly under severe weather conditions. The SBS block copolymers are included in the list of most commonly utilized modifiers for this particular application. Currently, it is confirmed that SEBS behaves as the better modifier as compared to SBS in enhancing the asphalts rutting resistance because of the double bond saturation due to which the SEBS is stiffer than SBS (Elseifi et al., 2003; Roland, 2013). Styrenic block copolymers can be compounded in order to yield materials that improve feel, appearance, and grip in applications like toys, personal hygiene, packaging, and automotive.

1.7.2. Thermoplastic Elastomers Centered on Polyamide

COPAs (thermoplastic polyamide elastomers) belong to the class of segmented block copolymers. They comprise of multiblock copolymers structure with repeating soft and hard segments. Generally, hard segments are the polyamides that act as near crosslinks decreasing the chain slippage and viscous flow of the copolymer, whereas the soft segments are polyesters or polyethers, which add to the extensibility and flexibility of elastomers. Both of the segments are associated with amide linkages. The significant members of COPAs are polyetheresteramides (PEEAs), (polyesteramides (PEAs), polycarbonate esteramides (PCEAs), and polyether-block-amides (PE-b-As).

The properties of thermoplastic polyamide elastomers (TPE-A) might change according to factors as the ratio of the soft and hard segments in a copolymer, molecular weight distribution, chemical composition, the thermal history of a sample, and the technique of preparation. The hard segment governs the extent of crystallinity, mechanical strength, and crystalline melting point whereas the soft segments control the hydrolytic stability, chemical resistance, low-temperature flexibility, and thermal-oxidative stability.

Majority of the COPAs display higher resistance to raised temperature as compared to any other commercial TPEs. They are more resistant to long-term heat aging without any addition of heat stabilizers. The abrasion resistance of TPE-A is analogous to that of the TPUs. COPAs has outstanding abrasion resistance and is similar to that of the TPUs. The hardness of COPAs is in the range of Shore 80A-Shore 70D by changing the content of soft and hard segments (Brydson, 1995). The better insulation properties of TPE-A make them appropriate for jacketing and low voltage applications. Other areas of application for this material consist of drive belt and conveyor, footwear like sports shoes and ski boots, automotive applications, hot melt adhesives, electronics, control modifiers for engineering thermoplastics, and powder coatings for the metals (Rodriguez-Galan et al., 2011).

Polyester amides institute a peculiar of the biodegradable family, because of the existence of both amide and ester groups that guarantees degradability. The biodegradable polymers are getting great consideration and are presently being developed for several biomedical applications like governed drug delivery systems, tissue engineering, and hydrogels (Sadhan and White, 2001; Cheng et al., 2011).

1.7.3. Thermoplastic Polyether Ester Elastomers

The COPEs (polyether ester elastomers or copolyesters) comprise a sequence of soft and hard segments. The blocks having high melting points are created by the segments of crystalline polyester, which are proficient in crystallization, and then the rubbery soft segments are created by the segments of an amorphous polyether having a comparatively low Tg (glass transition temperature). At advantageous service temperature, the blocks of polyester form the crystalline domains entrenched in rubbery polyether constant phase. The crystalline domains behave like physical crosslinks. At higher temperatures, the breakdown of crystallites occurs to produce a polymer melt, therefore simplifying the processing of thermoplastic.

COPEs are well-thought-out engineering TPEs due to their uncommon combination of elasticity, dynamic properties, and strength. They have a broad valuable temperature range amongst the Tg (nearly –50°C) and melting point (nearly 200°C) (Polacco et al., 2006; Jasso et al., 2015). These materials are normally elastic, but the recoverable elasticity is restricted to low strains. They also have outstanding dynamic performance and exhibit resistance to creep. The copolyesters can be utilized in the electrical uses for voltages around 600 Volts and less. They are resistant to aromatic and aliphatics hydrocarbons, oils, alcohols, esters, hydraulic fluids, and ketones.

1.7.4. Polyolefin-Centered Thermoplastic Elastomers

TPOs (Polyolefin thermoplastic elastomers) are a significant part of the TPEs, which comprise amorphous elastomeric and polyolefin semi-crystalline thermoplastic components. TPOs have a co-continuous system of phases with hard phases giving the strength and flexibility is provided by the soft phase. The two most significant processing techniques of TPOs are extrusion and injection molding. Other methods of processing include calendaring, negative thermoforming, blow molding, and thermoforming. TPOs ingredients usually include EPM (ethylene-propylene random copolymer), isotactic polypropylene (iPP), and the addition of additives and fillers (Zhao et al., 2016; Perju et al., 2020).

TPOs share the fundamental features of possessing the elastomeric properties with all of the TPEs, yet they process just like the thermoplastic material. They are accessible in the range of 60 Shore A-70 Shore D. The flexural modulus of TPE can vary from 1,000–2,50,000 psi. TPOs having high contents of elastomers are very rubbery, having a high elongation at the break values, whereas TPOs having high contents of polyolefin experience produce at the low elongation. Most of the polyolefin TPEs are resilient to ozone, unaffected by the aqueous solutions of chemicals or water. TPOs are outstanding electrical insulating materials having a dielectric strength of around 500 V/mil and volume resistivity at around 23°C and 50% RH is 1.6 9 1016 (Mcnamara et al., 2010).

Hard TPOs compounds are often utilized for exterior fascia or injection molded automotive interior. Soft TPOs compounds can normally be extruded into the sheet and then thermoformed for a large part of automotive like interior skin (Sureshkumar et al., 2010).

1.7.5. Thermoplastic Elastomers Made by Dynamic Vulcanization

Dynamic vulcanization is an extensively utilized method to make TPEs, including fully or partially crosslinked particles of an elastomer in the melt-processable thermoplastic matrix. TPEs prepared by this technique are known as thermoplastic vulcanizates (TPVs). The difference between the TPVs and TPOs is that the polyolefin and elastomeric phases in the TPOs are co-continuous whereas in TPVs, the elastomeric phase is discontinuous and the continuous polyolefin phase. A 2-D representation of TPVs and TPOs are exhibited in Figure 1.17.

TPVs can be processed with the help of fabrication and processing techniques usual to thermoplastics, like extrusion, compression molding, injection molding, blow molding, calendaring, extrusion foaming, and thermoforming (Soares et al., 2008). The commercialized TPVs are usually centered on a dynamically vulcanized rubber blend of EPDM with the polyolefin resin. Though, it is feasible to prepare TPVs with a multiplicity of elastomers and thermoplastics. Some of the utilized are polyolefins and diene rubbers, halobutyl, and butyl rubbers and polyacrylate rubber polyolefins and polyolefins (Passador et al., 2008), and polyvinyl chloride (PVC) and butadiene-acrylonitrile rubber (Ivan, 1991).

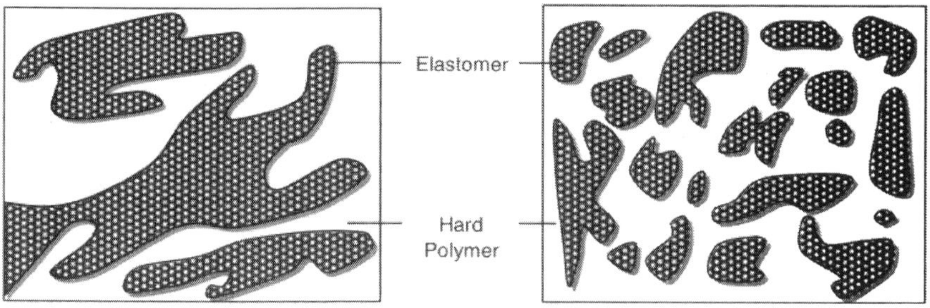

Figure 1.17. Growth of morphology in the thermoplastic vulcanizates from the co-continuous phase to isolated phase.

Source: https://www.researchgate.net/publication/244063716_Morphology_development_in_Thermoplastic_Vulcanizates_TPV_Dispersion_MechanismS_of_a_Pre-Crosslinked_EPDM_phase.

The ratio of thermoplastic and elastomeric components in a system is critical for the chief physical properties (modulus at 100% elongation, Young modulus, etc.), and its impact on the final properties (elongation at break, tensile strength, resilience, and tear strength, etc.). There are three decisive features to acquire TPVs having a better balance of properties: DC (wetting surface tensions) of the thermoplastic and elastomeric components, Wc (fraction of crystallinity) of a thermoplastic, and Nc (critical entanglement spacing) of the macromolecules of elastomer (Coran et al., 1982). The reduction of both elongation at break and tensile strength has been discovered with an upsurge in NC and D_C and a reduction in W_C (Lame, 2010).

1.8. ELASTOMER-FILLER COMPOSITIONS

The formulations of elastomer needed for applications necessitating extended performance comprise fillers, the most usual being the CB. Fillers like silica, talc, calcium carbonate, and kaolin are often utilized. Currently, organomodified clays, carbon nanotubes (CNTs), and nanoparticulate synthetic inorganic compounds have been presented. Filler elastomers display changes beyond those anticipated from the theories of volume fraction and considerable has been discussed regarding filler distribution and the interactions. The fractal nanostructure was utilized to understand heterogeneous deformation where the filler aggregates change with upsurges in strain (Suphadon et al., 2010). NR and SBR subjected to the large pre-strains of 100% exhibited loss modulus and storage behavior free from the pre-strain when normally occupied with 25 pph of CB whereas at higher content of filler to 50 pph the modulus of loss augmented with pre-strain. The impact of prestrain on the higher CB occupied elastomers was understood as because of the molecular slippage at filler interface (Stöckelhuber et al., 2010).

Dispersibility of the nano-fillers in elastomers depends upon the surface treatments in order to reduce filler-filler interactions whereas increasing compatibility between elastomer and filler. Comparative interaction between elastomer and filler has been assessed from polarity and surface energies and consequent work of adhesion (Sosson et al., 2010).

The existence of the filler aggregates and the creation of the filler agglomerates are factors time and strain dependent properties. The agglomerates can be interrupted by upsurging strain, although they might reform over time. Elastomer entangled inside filler agglomerates upsurges the operational filler volume fraction as the entangled elastomer is restrained and doesn't add to the elastic response till the filler agglomerate is disturbed at the larger deformations. The restriction of elastomer response by the filler has been defined as because of the tightly bound interphase accredited to the hydrodynamic effect that upsurges the operational filler volume fraction (Merabia et al., 2010).

Variations of the filled-elastomer response due to strain and over time amongst strains have been observed long described as Mullins and Payne effects. Much consideration to the understanding of the effects has currently occurred due to progress in the instrumentation and curiosity in the elastomer nanocomposites. Plastic deformation because of the clusters is overlaid on the elastic response (Lewicki et al., 2010).

The composites of nanofiller-elastomer display increased thermal stability accredited to binding of the elastomer molecules to nanofiller and the crookedness of the way for diffusion of the volatile degradation products. Polyhedral oligomeric silsesquioxanes (POSS) are the single-molecule nanofillers and have gained much current attention; for instance, they have been utilized as a part of polyol components in the PU elastomers. Thermogravimetry exhibited that POSS-PU hybrids have augmented thermal stability because of the impact of POSS on the other degradation mechanism (Kuila et al., 2009; Sun et al., 2009). The mechanical properties were increased whereas fire retardance and thermal degradation resistance.

The particles of nano-silica are now incorporated in numerous elastomeric products, the main example being polysiloxane, though silica has been discovered to improve hydrocarbon elastomers. Silica was incorporated in the solution polymerized poly(butadiene-co-styrene) and made the non-linear stress relaxation because of the juxtaposition of the elastomer network and the silica network that contributed to non-linearity (Şerbescu and Saalwächter, 2009). The nano-fillers like silica comprise non-dispersible aggregates that normally flocculate into the larger agglomerates prevailing as fractal arrays inside the elastomer. Agglomerates can be interrupted by the elastomer strain and reshape upon relaxation. Agglomerates might comprise of occluded elastomer consequential in an exaggeratedly high obvious filler volume fraction, the phenomenon called the Payne effect. Around 25% of w/w smoked silica was distributed in a poly(dimethylsiloxane) that was often discovered by NMR to have the two-phase morphology. Microscopic network variations happened with the storage time of poly(dimethylsiloxane)–silica composites (Mathai and Thomas, 2005; Wu et al., 2010).

1.9. ELASTOMER BLENDS

The polymer blends are categorized are incompatible, miscible, compatible. Polymer blends appropriate for application should be compatible although they might be miscible frequently miscibility isn't anticipated as the properties are averaged. Thermoplastic PU and SBS centered upon polyester or polyether offered transparent blends because of the narrow-distributed phase size of the particle and comparable refractive index of amalgamated elastomers. The blends of SBS-TPU exhibited enhanced thermal resistance, damping at the high frequency, and mechanical properties (Mathai and Thomas, 2005). Membranes made from blends of epoxidized NR and nitrile rubber has been made and there, miscibility, morphology, viscoelastic, and mechanical properties have usually been measured. The blend composition inclined homogeneity, viscoelasticity, and microdomain structure (Stephen et al., 2003).

The blends were made from NR with the carboxylated poly(butadiene co-styrene) rubber and temperature and composition property dependences were examined. 2 glass transitions were discovered and the other data was consistent with the uniform 2-phase system where viscoelasticity and morphology were governed by composition (Tang et al., 2003).

1.10. INTERPENETRATING ELASTOMER BLENDS (IPN)

IPN (Interpenetrating polymer/elastomer networks) can be made by instantaneous and sequential crosslinking or polymerization to provide the co-continuous network of chemically distinct polymers (Figure 1.18). The uncross-linked, blended pre-polymers or elastomer polymers must be primarily miscible to permit the interpenetrating network in order to create at a molecular level. As the distinct crosslinking reactions continue, entropically, the polymers phase isolate creating the co-continuous networks. The polymers might be elastomers, or one of them an elastomer and the second a polymer in order to improve toughness. For instance, the acrylate-altered PU was interpenetrated through an unsaturated polyester resin in order to form the gradient IPN (Zhao et al., 2016; Perju et al., 2020).

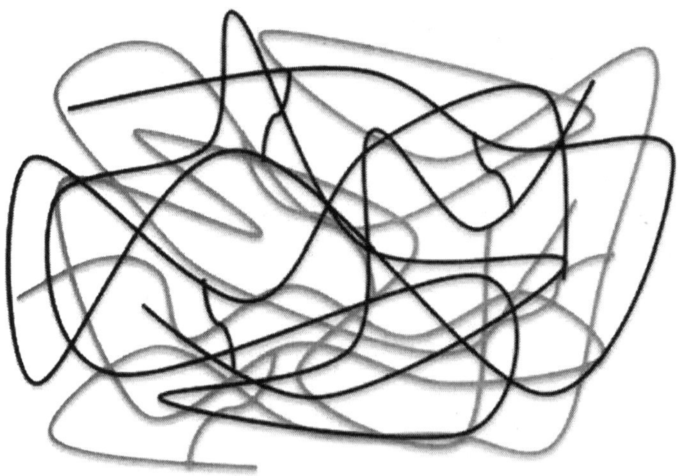

Figure 1.18. Graphic structure for the interpenetrating polymer blend.

Source: https://www.researchgate.net/figure/8-Schematic-structure-for-an-interpenetrating-polymer-blend_fig16_260751886.

The interpenetrated domains were discovered to be of the nanometer dimension, having glass transitions associated such that the interpenetrating elastomer blends changed from elastomeric to brittle (Guo et al., 2001).

The blends of a poly(oxypropylene)-kind epoxy resin and the poly(butadiene-b-styrene) were made and instantaneously cured in order to form the network with three glass transitions, because of two anticipated phases from a poly(butadiene-b-styrene) and the third epoxy phase (Song et al., 1999). The interphases in the polybutadiene-NR blend, the poly(vinyl acetate)-poly(methyl methacrylate) structured latex film, the poly(vinyl acetate)-poly(epichlorohydrin) bilayer film, and the polystyrene-PU and poly(ethyl methacrylate)–PU interpenetrating networks of polymer were examined with the help of modulated temperature differential scanning calorimetry (mTDSC). The outcomes of mT-DSC recognized glass transitions of separate components and the interphases (Shanks, 2013).

1.11. FORMULATION AND COMPOUNDING

The formulations of elastomers comprise additional components like curing agents or crosslinking, accelerators, and initiators for curing, stabilizers like antioxidants, plasticizers, processing lubricants or extenders, and the fillers like silica and CB (Kumar et al., 2010). Elastomer-centered adhesives are utilized as connection adhesives, for instance. The initial adhesion needs the addition of the tackifier with phenol-formaldehyde, and rosin as instances. The brominated isobutylene-co-p-methyl styrene rubber pacified with an amalgamated hydrocarbon resin and the maleated resin tackifier of hydrocarbon has been discovered to lead to enhanced performance during de-bonding, and bonding measured utilizing the peal test and depended on the majority of viscoelastic properties (Meng and Hu, 2009).

1.12. SHAPE-MEMORY POLYMERS

These are elastomers in a temperature range where they are generally deformed from the equilibrium shape, though, at surrounding temperatures they should preserve their deformed shape. At the ambient temperatures, they should be quite below the glass transition of elastomeric continuous phase in order to preserve the deformed shape for a very long time till the memory is triggered by heating. The Tg (glass transition temperature) must be suitably above any possible ambient temperatures, although normally adequately low, so that the elastomeric behavior can be triggered utilizing moderate heat like hot water and hot air from the heated air gun like a hairdryer and some other handy heating devices. When warmed into the elastomeric system shape-memory materials must have exact elastomeric behavior that is they must have reversible and rapid deformations without any creep, which is with minimal viscoelasticity and irreversible since the instant return to the original shape is needed (Ratna and Karger-Kocsis, 2008; Liu et al., 2009). PUs are usual shape-memory materials since they display elastomeric properties over the temperature ranges governed by characteristic and structure hydrogen bonding offers limitations on molecular motions and conformational change. The shape-memory reaction might be activated electrically instead of the use of mechanical force (Bagdi et al., 2011).

1.13. SUMMARY AND THE FUTURE TRENDS

The trend with novel polymers is towards specialty applications where advanced and unique properties are required. The necessities are met through enhanced control of the molecular structure, formulation, and copolymerization of prevailing polymer kinds. Elastomers trail this inclination with novel polymerization methods and catalysts to govern tacticity, molar mass distribution, and molar mass. Formulation invention has come from the blends of polymer that are well-suited, although not miscible normally when averaging of the properties would take place. Current publication frequencies exhibit that the nanocomposites are elastomers with the quickest development. The elastomers nanocomposites have been substantial in that silicas and CBs utilized in the traditional industry of rubber are the nanoparticulate materials. Improved energy damping and elastomeric properties are major areas of novel developments. Additionally, the resistant elastomers are usually in demand for thermal resistance, chemical resistance, radiation resistance, weathering resistance, and wear/abrasion resistance (Cvek et al., 2017).

The chapter has defined the properties of the elastomer and that such kind of materials should be polymeric. The elastomers are tangential polymers preferably having high molar mass in order to form the random coils, and these coils can usually be prolonged towards the entropically unbalanced linear conformation. These polymers have backed considerably to a comprehension of the behavior of macromolecules. The structural refinement is needed in practice, at first, some verge of crosslinking avert flow therefore limiting molecules to the reversible uncoiling. The further modification comprises chemical structures in order to give resistance to the severe environments, functionality to postpone an elastic response in order to improve energy damping and toughness. Other components active mechanically include blended polymers and fillers that might be other elastomers, compatible or miscible polymers. The elastomers find utilization in the large array of produces where their part might be evident or they might contribute a significant but concealed function. The NR was one of the initial polymers utilized and important to even present economies. New nanofillers and blends now refine the properties. The diversified elastomer groups like PUs, fluorocarbon polymers, and polyacrylates contribute exclusive property combinations to diversified and specialized products of high technology. The inclination towards biomaterials signs the return to the significance of NR, though there is a chance for the development of novel hybrid bioelastomers (Gonzaga et al., 2009).

REFERENCES

1. Abdeen, Z. I., (2017). Polyurethane rubber-based nanoblends: Preparation, characterization and applications. In: *Rubber Nano-Blends* (Vol. 1, pp. 89–103). Springer, Cham.
2. Adam, C., Lacoste, J., & Lemaire, J., (1989). Photo-oxidation of elastomeric materials. Part 1: Photo-oxidation of polybutadienes. *Polymer Degradation and Stability, 24*(3), 185–200.
3. Akram, N., Zia, K. M., Sattar, R., Tabassum, S., & Saeed, M., (2019). Thermomechanical investigation of hydroxyl-terminated polybutadiene-based linear polyurethane elastomers. *Journal of Applied Polymer Science, 136*(13), 47289.
4. Ayoub, G., Zaïri, F., Naït-Abdelaziz, M., & Gloaguen, J. M., (2011). Modeling the low-cycle fatigue behavior of visco-hyper elastic elastomeric materials using a new network alteration theory: Application to styrene-butadiene rubber. *Journal of the Mechanics and Physics of Solids, 59*(2), 473–495.
5. Bagdi, K., Molnár, K., Sajo, I., & Pukánszky, B., (2011). Specific interactions, structure and properties in segmented polyurethane elastomers. *Express Polymer Letters, 5*(5).
6. Blackwell, J., & Gardner, K. H., (1979). Structure of the hard segments in polyurethane elastomers. *Polymer, 20*(1), 13–17.
7. Bokobza, L., & Kolodziej, M., (2006). On the use of carbon nanotubes as reinforcing fillers for elastomeric materials. *Polymer International, 55*(9), 1090–1098.
8. Borgmann, S., Schulte, A., Neugebauer, S., & Schuhmann, W., (2011). Amperometric biosensors. *Advances in Electrochemical Science and Engineering, 2, 1–33.*
9. Brackbill, C. R., Lesieutre, G. A., Smith, E. C., & Govindswamy, K., (1996). Thermomechanical modeling of elastomeric materials. *Smart Materials and Structures, 5*(5), 529.
10. Breiner, J. M., Mark, J. E., & Beaucage, G., (1999). Dependence of silica particle sizes on network chain lengths, silica contents, and catalyst concentrations in in situ-reinforced polysiloxane elastomers. *Journal of Polymer Science Part B: Polymer Physics, 37*(13), 1421–1427.

11. Brown, D. W., Lowry, R. E., & Smith, L. E., (1980). Kinetics of hydrolytic aging of polyester urethane elastomers. *Macromolecules, 13*(2), 248–252.
12. Brydson, J. A., (1995). *Thermoplastic Elastomers: Properties and Applications* (Vol. 81, pp. 1–33). Smithers Rapra Publishing.
13. Buckley, C. P., Prisacariu, C., & Martin, C., (2010). Elasticity and inelasticity of thermoplastic polyurethane elastomers: Sensitivity to chemical and physical structure. *Polymer, 51*(14), 3213–3224.
14. Cantournet, S., Desmorat, R., & Besson, J., (2009). Mullins effect and cyclic stress softening of filled elastomers by internal sliding and friction thermodynamics model. *International Journal of Solids and Structures, 46*(11–12), 2255–2264.
15. Cella, R. J., (1973). Morphology of segmented polyester thermoplastic elastomers. In: *Journal of Polymer Science: Polymer Symposia* (Vol. 42, No. 2, pp. 727–740). New York: Wiley Subscription Services, Inc., A Wiley Company.
16. Chazeau, L., Brown, J. D., Yanyo, L. C., & Sternstein, S. S., (2000). Modulus recovery kinetics and other insights into the Payne effect for filled elastomers. *Polymer Composites, 21*(2), 202–222.
17. Cheng, H., Hill, P. S., Siegwart, D. J., Vacanti, N., Lytton-Jean, A. K., Cho, S. W., & Anderson, D. G., (2011). A novel family of biodegradable poly (ester amide) elastomers. *Advanced Materials, 23*(12), H95–H100.
18. Cheng, Q., Wang, Y., Su, L., Wang, H., Zhu, G., & Yu, W., (2019). Wide-spectrum manipulation of triboelectrification-induced electroluminescence by long afterglow phosphors in elastomeric zinc sulfide composites. *Journal of Materials Chemistry C, 7*(15), 4567–4572.
19. Chester, S. A., & Anand, L., (2010). A coupled theory of fluid permeation and large deformations for elastomeric materials. *Journal of the Mechanics and Physics of Solids, 58*(11), 1879–1906.
20. Chiang, F. C., Cai, S., Jin, A. K. S., Bauer, S., & Suo, Z., (2012). Model of dissipative dielectric elastomers. *Journal of Applied Physics, 111*(3), 034102.
21. Coran, A. Y., Patel, R. P., & Williams, D., (1982). Rubber-thermoplastic compositions. Part V: Selecting polymers for thermoplastic vulcanizates. *Rubber Chemistry and Technology, 55*(1), 116–136.

22. Courval, G., & Gray, D. G., (1975). The effect of surface adsorption on gas chromatographic measurements near polymer melting transitions. *Macromolecules, 8*(3), 326–331.

23. Cvek, M., Moucka, R., Sedlacik, M., & Pavlinek, V., (2017). Electromagnetic, magnetorheological and stability properties of polysiloxane elastomers based on silane-modified carbonyl iron particles with enhanced wettability. *Smart Materials and Structures, 26*(10), 105003.

24. Datta, S., (2004). Special-purpose elastomers. *Brendan Rodgers. Rubber Compounding: Chemistry and Applications, 1*, 105–132. Marcel Dekker Inc, New York.

25. Elias, H. G., Meyer, T., Keurentjes, J., Odian, G., Buback, M., Herk, A. M. V., & Hutchinson, R. A., (2006). *Macromolecular Symposia Related Titles, 1*, 1–33.

26. Elias, H. G., Munk, P., & Aminabhavi, T. M., (1990). Introduction to macromolecular science. *Journal of Colloid and Interface Science, 137*(1), 311–311.

27. Elseifi, M. A., Flintsch, G. W., & Al-Qadi, I. L., (2003). Quantitative effect of elastomeric modification on binder performance at intermediate and high temperatures. *Journal of Materials in Civil Engineering, 15*(1), 32–40.

28. Flory, P. J., (1953). *Principles of Polymer Chemistry* (Vol. 1, pp. 432–494). Cornell University Press.

29. Gamlin, C., Markovic, M., Dutta, N., Choudhury, N., & Matisons, J., (2000). Structural effects on the decomposition kinetics of EPDM elastomers by high-resolution TGA and modulated TGA. *Journal of Thermal Analysis and Calorimetry, 59*(1, 2), 319–336.

30. Gonzaga, F., Yu, G., & Brook, M. A., (2009). Polysiloxane elastomers via room temperature, metal-free click chemistry. *Macromolecules, 42*(23), 9220–9224.

31. Goor, O. J., Brouns, J. E., & Dankers, P. Y., (2017). Introduction of anti-fouling coatings at the surface of supramolecular elastomeric materials via post-modification of reactive supramolecular additives. *Polymer Chemistry, 8*(34), 5228–5238.

32. Guenthner, R. A., & Stivers, D. A., (1984). *U.S. Patent No. 4,446,270* (pp. 1–20). Washington, DC: U.S. Patent and Trademark Office.

33. Gunatillake, P. A., Martin, D. J., Meijs, G. F., McCarthy, S. J., & Adhikari, R., (2003). Designing biostable polyurethane elastomers for biomedical implants. *Australian Journal of Chemistry, 56*(6), 545–557.
34. Guo, Q., Figueiredo, P., Thomann, R., & Gronski, W., (2001). Phase behavior, morphology and interfacial structure in thermoset/thermoplastic elastomer blends of poly(propylene glycol)-type epoxy resin and polystyrene-b-polybutadiene. *Polymer, 42*(26), 10101–10110.
35. Haupt, P., & Sedlan, K., (2001). Viscoplasticity of elastomeric materials: Experimental facts and constitutive modeling. *Archive of Applied Mechanics, 71*(2, 3), 89–109.
36. Hong, Y., Guan, J., Fujimoto, K. L., Hashizume, R., Pelinescu, A. L., & Wagner, W. R., (2010). Tailoring the degradation kinetics of poly (ester carbonate urethane) urea thermoplastic elastomers for tissue engineering scaffolds. *Biomaterials, 31*(15), 4249–4258.
37. Ivan, G., (1991). Dynamic vulcanization: An accessible way to thermoplastic elastomers. *Iran. J. Polym. Sci. Technol. (Engl. Ed.), 2*(1), 3–11.
38. Jasso, M., Hampl, R., Vacin, O., Bakos, D., Stastna, J., & Zanzotto, L., (2015). Rheology of conventional asphalt modified with SBS, elvaloy and polyphosphoric acid. *Fuel Processing Technology, 140*, 172–179.
39. Jeong, S. M., Song, S., Kim, H., Joo, K. I., & Takezoe, H., (2016). Mechanoluminescence color conversion by spontaneous fluorescent-dye-diffusion in elastomeric zinc sulfide composite. *Advanced Functional Materials, 26*(27), 4848–4858.
40. Kim, I., Ahn, J. T., Ha, C. S., Yang, C. S., & Park, I., (2003). Polymerization of propylene oxide by using double metal cyanide catalysts and the application to polyurethane elastomer. *Polymer, 44*(11), 3417–3428.
41. Klingender, R. C., (2008). *Handbook of Specialty Elastomers* (Vol. 1, pp. 1–22). CRC Press.
42. Krol, P., (2007). Synthesis methods, chemical structures and phase structures of linear polyurethanes. Properties and applications of linear polyurethanes in polyurethane elastomers, copolymers and ionomers. *Progress in Materials Science, 52*(6), 915–1015.
43. Kuila, T., Srivastava, S. K., & Bhowmick, A. K., (2009). Rubber/LDH nanocomposites by solution blending. *Journal of Applied Polymer Science, 111*(2), 635–641.

44. Kumar, D., & Sarangi, S., (2019). Electro-mechanical instability modeling in elastomeric actuators: A second law of thermodynamics-based approach. *Soft Materials, 17*(3), 308–320.
45. Kumar, K. D., Tsou, A. H., & Bhowmick, A. K., (2010). Interplay between bulk viscoelasticity and surface energy in autohesive tack of rubber-tackifier blends. *Journal of Polymer Science Part B: Polymer Physics, 48*(9), 972–982.
46. Lame, O., (2010). Does fractal nanostructure of filled rubber lead to fractal deformations? *In situ* measurements of strain heterogeneities by AFM. *Macromolecules, 43*(13), 5881–5887.
47. LeBaron, P. C., & Pinnavaia, T. J., (2001). Clay nanolayer reinforcement of a silicone elastomer. *Chemistry of Materials, 13*(10), 3760–3765.
48. Lewicki, J. P., Pielichowski, K., De La Croix, P. T., Janowski, B., Todd, D., & Liggat, J. J., (2010). Thermal degradation studies of polyurethane/POSS nanohybrid elastomers. *Polymer Degradation and Stability, 95*(6), 1099–1105.
49. Li, Q., Dimaki, A., Popov, M., Psakhie, S. G., & Popov, V. L., (2014). Kinetics of the coefficient of friction of elastomers. *Scientific Reports, 4*, 5795.
50. Lin, Y., Hansen, H. R., Brittain, W. J., & Craig, S. L., (2019). Strain-dependent kinetics in the cis-to-trans isomerization of azobenzene in bulk elastomers. *The Journal of Physical Chemistry B, 123*(40), 8492–8498.
51. Liu, L., Liu, Y., Li, B., Yang, K., Li, T., & Leng, J., (2011). Thermo-electro-mechanical instability of dielectric elastomers. *Smart Materials and Structures, 20*(7), 075004.
52. Liu, Y., Lv, H., Lan, X., Leng, J., & Du, S., (2009). Review of electro-active shape-memory polymer composite. *Composites Science and Technology, 69*(13), 2064–2068.
53. Logothetis, A. L., (1989). Chemistry of fluorocarbon elastomers. *Progress in Polymer Science, 14*(2), 251–296.
54. MacLachlan, J. D., (1978). Fluorocarbon elastomers: A technical review. *Polymer-Plastics Technology and Engineering, 11*(1), 41–53.
55. Mahaling, R. N., Jana, G. K., Das, C. K., Jeong, H., & Ha, C. S., (2005). Carboxylated nitrile elastomer/filler nanocomposite: Effect of silica nanofiller in thermal, dynamic mechanical behavior, and interfacial adhesion. *Macromolecular Research, 13*(4), 306–313.

56. Masamoto, J., & Kubo, K., (1996). Elastomer-toughened poly (phenylene sulfide). *Polymer Engineering and Science, 36*(2), 265–270.
57. Mathai, A. E., & Thomas, S., (2005). Morphology, mechanical and viscoelastic properties of nitrile rubber/epoxidized natural rubber blends. *Journal of Applied Polymer Science, 97*(4), 1561–1573.
58. Mazurek, M. M., Tomczyk, K., & Rokicki, G., (2014). PET wastes utilization in the synthesis of aliphatic-aromatic polyurethane elastomers. *Polymers for Advanced Technologies, 25*(11), 1273–1284.
59. McNamara, J. J., Wood, M. G., & Dong, Y., (2010). *U.S. Patent No. 7,642,320* (Vol. 1, pp. 1–22). Washington, DC: U.S. Patent and Trademark Office.
60. Meng, Q., & Hu, J., (2009). A review of shape memory polymer composites and blends. *Composites Part A: Applied Science and Manufacturing, 40*(11), 1661–1672.
61. Merabia, S., Sotta, P., & Long, D. R., (2010). Unique plastic and recovery behavior of nanofilled elastomers and thermoplastic elastomers (Payne and Mullins effects). *Journal of Polymer Science Part B: Polymer Physics, 48*(13), 1495–1508.
62. Mooney, M., (1948). The thermodynamics of a strained elastomer. I. General analysis. *Journal of Applied Physics, 19*(5), 434–444.
63. Myung, D., Waters, D., Wiseman, M., Duhamel, P. E., Noolandi, J., Ta, C. N., & Frank, C. W., (2008). Progress in the development of interpenetrating polymer network hydrogels. *Polymers for Advanced Technologies, 19*(6), 647–657.
64. Pandey, A., & Mehtra, A., (2014). Comparative study of dimensional stability and accuracy of various elastomeric materials. *IOSR Journal of Dental and Medical Sciences (IOSR-JDMS), 13*(3), 40–45.
65. Passador, F. R., Rodolfo, A., & Pessan, I. A., (2008). Dynamic vulcanization of PVC/NBR blends. In: *Proceedings of the Polymer Processing Society 24th Annual Meeting. The Polymer Processing Society* (Vol. 1, pp. 1–22). Salermo.
66. Paul, D. R., & Mark, J. E., (2010). Fillers for polysiloxane (silicone) elastomers. *Progress in Polymer Science, 35*(7), 893–901.
67. Perju, E., Ko, Y. S., Dünki, S. J., & Opris, D. M., (2020). Increased electromechanical sensitivity of polysiloxane elastomers by chemical

modification with thioacetic groups. *Materials and Design, 186*, 108319.

68. Petrović, Z. S., & Ferguson, J., (1991). Polyurethane elastomers. *Progress in Polymer Science, 16*(5), 695–836.
69. Polacco, G., Muscente, A., Biondi, D., & Santini, S., (2006). Effect of composition on the properties of SEBS modified asphalts. *European Polymer Journal, 42*(5), 1113–1121.
70. Polacco, G., Stastna, J., Biondi, D., & Zanzotto, L., (2006). Relation between polymer architecture and nonlinear viscoelastic behavior of modified asphalts. *Current Opinion in Colloid and Interface Science, 11*(4), 230–245.
71. Polmanteer, K. E., & Hunter, M. J., (1959). Polymer composition versus low-temperature characteristics of polysiloxane elastomers. *Journal of Applied Polymer Science, 1*(1), 3–10.
72. Promma, N., Raka, B., Grediac, M., Toussaint, E., Le Cam, J. B., Balandraud, X., & Hild, F., (2009). Application of the virtual fields method to mechanical characterization of elastomeric materials. *International Journal of Solids and Structures, 46*(3, 4), 698–715.
73. Qi, H. J., & Boyce, M. C., (2004). Constitutive model for stretch-induced softening of the stress-stretch behavior of elastomeric materials. *Journal of the Mechanics and Physics of Solids, 52*(10), 2187–2205.
74. Qi, H. J., Joyce, K., & Boyce, M. C., (2003). Durometer hardness and the stress-strain behavior of elastomeric materials. *Rubber Chemistry and Technology, 76*(2), 419–435.
75. Ratna, D., & Karger-Kocsis, J., (2008). Recent advances in shape memory polymers and composites: A review. *Journal of Materials Science, 43*(1), 254–269.
76. Rodas, C. O., Zaïri, F., & Naït-Abdelaziz, M., (2014). A finite strain thermo-viscoelastic constitutive model to describe the self-heating in elastomeric materials during low-cycle fatigue. *Journal of the Mechanics and Physics of Solids, 64*, 396–410.
77. Rodriguez-Galan, A., Franco, L., & Puiggali, J., (2011). Degradable poly (ester amide)s for biomedical applications. *Polymers, 3*(1), 65–99.
78. Rogulska, M., & Kultys, A., (2016). Aliphatic polycarbonate-based thermoplastic polyurethane elastomers containing diphenyl sulfide units. *Journal of Thermal Analysis and Calorimetry, 126*(1), 225–243.

79. Roland, C. M., (2013). Immiscible rubber blends. In: *Advances in Elastomers I* (Vol. 1, pp. 167–181). Springer, Berlin, Heidelberg.
80. Sadhan, K., & White, J. R., (2001). *De Rubber Technologist's Handbook*. Rapra Technology Limited: Shawbury, Shrewsbury, Shropshire, SY4 4NR, UK.
81. Sare, I. R., Mardel, J. I., & Hill, A. J., (2001). Wear-resistant metallic and elastomeric materials in the mining and mineral processing industries: An overview. *Wear, 250*(1–12), 1–10.
82. Şerbescu, A., & Saalwächter, K., (2009). Particle-induced network formation in linear PDMS filled with silica. *Polymer, 50*(23), 5434–5442.
83. Shanks, R. A., (2013). General-purpose elastomers: Structure, chemistry, physics and performance. In: *Advances in Elastomers I* (Vol. 1, pp. 11–45). Springer, Berlin, Heidelberg.
84. Sivakumar, A., & Gopal, R., (2011). *Halogen-Free Synthetic Elastomer Blend to Meet Properties of Fuel Hose Outer Cover (Return Line) Application* (No. 2011–01–2233, pp. 1–33). SAE Technical Paper.
85. Soares, B. G., Santos, D. M., & Sirqueira, A. S., (2008). A novel thermoplastic elastomer based on dynamically vulcanized polypropylene/acrylic rubber blends. *Express Polymer Letters, 2*(8), 602–613.
86. Song, M., Hourston, D., Reading, M., Pollock, H., & Hammiche, A., (1999). Modulated differential scanning calorimetry analysis of interphases in multi-component polymer materials. *Journal of Thermal Analysis and Calorimetry, 56*(3), 991–1004.
87. Sonnenschein, M. F., Guillaudeu, S. J., Landes, B. G., & Wendt, B. L., (2010). Comparison of adipate and succinate polyesters in thermoplastic polyurethanes. *Polymer, 51*(16), 3685–3692.
88. Sosson, F., Belec, L., Chailan, J. F., Carriere, P., & Crespy, A., (2010). Highlight of a compensation effect between filler morphology and loading on dynamic properties of filled rubbers. *Journal of Applied Polymer Science, 117*(5), 2715–2723.
89. Sotayo, A., Bradley, D., Bather, M., Sareh, P., Oudjene, M., El-Houjeyri, I., & Guan, Z., (2020). Review of state of the art of dowel laminated timber members and densified wood materials as sustainable engineered wood products for construction and building applications. *Developments in the Built Environment, 1*, 100004.

90. Sperling, L. H., (2005). *Introduction to Physical Polymer Science* (Vol. 1, pp. 1–20). John Wiley & Sons.
91. Steffen, R., Meir, M., Rekstad, J., & Röder, B., (2018). Kinetics of degradation-induced polymer luminescence: A polyphenylene sulfide/elastomer blend under dry heat exposure. *Polymer, 136*, 71–83.
92. Stephen, R., Raju, K. V. S. N., Nair, S. V., Varghese, S., Oommen, Z., & Thomas, S., (2003). Mechanical and viscoelastic behavior of natural rubber and carboxylated styrene-butadiene rubber latex blends. *Journal of Applied Polymer Science, 88*(11), 2639–2648.
93. Stöckelhuber, K. W., Das, A., Jurk, R., & Heinrich, G., (2010). Contribution of Physico-chemical properties of interfaces on dispersibility, adhesion, and flocculation of filler particles in rubber. *Polymer, 51*(9), 1954–1963.
94. Sun, J., Li, H., Song, Y., Zheng, Q., He, L., & Yu, J., (2009). Nonlinear stress relaxation of silica filled solution-polymerized styrene-butadiene rubber compounds. *Journal of Applied Polymer Science, 112*(6), 3569–3574.
95. Suphadon, N., Thomas, A. G., & Busfield, J. J. C., (2010). The viscoelastic behavior of rubber under a small simple shear oscillation superimposed on a large pure shear. *Polymer Testing, 29*(4), 440–444.
96. Suphadon, N., Thomas, A. G., & Busfield, J. J. C., (2010). The viscoelastic behavior of rubber under a complex loading. II. The effect large strains and the incorporation of carbon black. *Journal of Applied Polymer Science, 117*(3), 1290–1297.
97. Sureshkumar, M. S., Filippi, S., Polacco, G., Kazatchkov, I., Stastna, J., & Zanzotto, L., (2010). Internal structure and linear viscoelastic properties of EVA/asphalt nanocomposites. *European Polymer Journal, 46*(4), 621–633.
98. Tang, D., Qin, C., Cai, W., & Zhao, L., (2003). Preparation, morphology, and mechanical properties of modified-PU/UPR graft-IPN nanocomposites with $BaTiO_3$ fiber. *Materials Chemistry and Physics, 82*(1), 73–77.
99. Tang, Z. P., & Gupta, Y. M., (1988). Shock-induced phase transformation in cadmium sulfide dispersed in an elastomer. *Journal of Applied Physics, 64*(4), 1827–1837.

100. Tikhomirov, S. G., Karmanova, O. V., Podvalny, S. L., Khvostov, A. A., & Karmanov, A. V., (2018). Research into kinetics of radiation destruction of elastomers. *Advanced Materials and Technologies,* (2), 1–33.
101. Ula, S. W., Traugutt, N. A., Volpe, R. H., Patel, R. R., Yu, K., & Yakacki, C. M., (2018). Liquid crystal elastomers: An introduction and review of emerging technologies. *Liquid Crystals Reviews, 6*(1), 78–107.
102. Ullman, R., (1986). Memory and lattice constraints in rubber elasticity. *Macromolecules, 19*(6), 1748–1753.
103. Van, A. G. J., (1950). Influence of structure of elastomers on their permeability to gases. *Journal of Polymer Science, 5*(3), 307–332.
104. Wang, B., Lu, H., & Kim, G. H., (2002). A damage model for the fatigue life of elastomeric materials. *Mechanics of Materials, 34*(8), 475–483.
105. Wang, H., Chen, X., Tian, Z., Jiang, Z., Yu, W., Ding, W., & Su, L., (2020). Efficient color manipulation of zinc sulfide-based mechanoluminescent elastomers for visualized sensing and anti-counterfeiting. *Journal of Luminescence, 228*, 117590.
106. Wang, R., Ma, J., Zhou, X., Wang, Z., Kang, H., Zhang, L., & Kulig, J., (2012). Design and preparation of a novel cross-linkable, high molecular weight, and bio-based elastomer by emulsion polymerization. *Macromolecules, 45*(17), 6830–6839.
107. Wang, W., Schlegel, R., White, B. T., Williams, K., Voyloy, D., Steren, C. A., & Mays, J., (2016). High temperature thermoplastic elastomers synthesized by living anionic polymerization in hydrocarbon solvent at room temperature. *Macromolecules, 49*(7), 2646–2655.
108. Wu, J. H., Li, C. H., Wu, Y. T., Leu, M. T., & Tsai, Y., (2010). Thermal resistance and dynamic damping properties of poly (styrene-butadiene-styrene)/thermoplastic polyurethane composites elastomer material. *Composites Science and Technology, 70*(8), 1258–1264.
109. Wündrich, K., (1984). A review of radiation resistance for plastic and elastomeric materials. *Radiation Physics and Chemistry (1977), 24*(5, 6), 503–510.
110. Zaharescu, T., Vilcu, R., & Podina, C., (1998). Some kinetic and thermodynamic aspects of thermal degradation of lightly stabilized elastomers. *Polymer Testing, 17*(8), 587–596.

111. Zhao, J., Xu, R., Luo, G., Wu, J., & Xia, H., (2016). A self-healing, re-moldable and biocompatible crosslinked polysiloxane elastomer. *Journal of Materials Chemistry B, 4*(5), 982–989.

112. Zhao, X., Koh, S. J. A., & Suo, Z., (2011). Nonequilibrium thermodynamics of dielectric elastomers. *International Journal of Applied Mechanics, 3*(02), 203–217.

CHAPTER 2

Classification of Elastomeric Materials

Contents

2.1. Introduction 52
2.2. Natural Rubber (NR) 55
2.3. Isoprene Rubber, Polyisoprene (IR) 58
2.4. Butadiene Rubber (BR), Polybutadiene 60
2.5. Styrene-Butadiene Rubber (SBR) 63
2.6. Butyl Rubbers 67
2.7. Nitrile Rubber, Nitrile-Butadiene Rubber, and Acrylonitrile Rubber (NBR) 69
2.8. Epichlorohydrin Rubbers 71
2.9. Ethylene-Propylene Rubbers 73
2.10. Chloroprene Rubber, Polychloroprene (CR) 76
2.11. Polyacrylate Rubbers (ACM) 77
2.12. Polyurethane Rubbers (AU, EU, PUR) 79
2.13. Fluorocarbon Rubbers (FKM, FPM) 82
2.14. Silicone Rubbers (Q) 85
2.15. Polysulfide Rubbers (T) 87
2.16. Ethylene-Vinyl Acetate Copolymer (EVA) 89
2.17. Polypropylene Oxide Rubbers, Polypropylene Oxide-Allyl Glycidyl Ether Copolymer (GPO) 90
2.18. Chlorinated Polyethylene (CM, CPE), Chlorosulfonated Polyethylene (CSM, CSPE) 91
2.19. Thermoplastic Elastomers (TPE) 93
2.20. Styrenic Thermoplastic Elastomers (TPE-S) 93
2.21. Elastomeric Alloys 95
2.22. Thermoplastic Urethane Elastomers (TPU, TPE-U) 97
2.23. Thermoplastics Polyester-Ether Elastomer (TPE-E) 99
2.24. Thermoplastic Polyamide Elastomers (TPE-A) 100
References 103

2.1. INTRODUCTION

Currently, there are numerous different kinds of elastomeric materials that could be used to enhance your application or product. The actual properties and characteristics the synthetic rubber material assumes would greatly rely on the chemicals utilized in production. These characteristics come right from the chemicals utilized. As a consequence, different elastomer kinds could range from being as tough as a softball or soft as a pillow. Due to this, it's dynamic you choose the best kinds of rubber materials to boost the performance of your application or product (Capps, 1989; Sanders et al., 2004).

There are two primary groups for elastomers, for example, unsaturated elastomers and saturated elastomers. Non-sulfur and sulfur vulcanization could cure unsaturated elastomers like butyl rubber, nitrile rubber, chloroprene rubber (CR), synthetic polyisoprene, polybutadiene, and others (Tobolsky et al., 1944; Andrews et al., 1946). Saturated elastomers like polyether block amides, polyether block amides, silicone rubber (SR), and polyacrylic rubber, ethylene-vinyl acetate, and much more have greater stability against heat, oxygen (O_2), ozone, and radiation. Sulfur vulcanization can't cure them. They have inferior reactivity and only react in a restricted number of situations under precise conditions (Shankar et al., 2007; Amin and Amin, 2011).

There are mainly two comprehensive categories into which the rubber kinds could be placed. These are Synthetic Rubber and Natural Rubber (NR). Occasionally vulcanized rubber is also taken to be a kind of rubber. Let's know about all these kinds of rubber (Eames et al., 1979; Gama et al., 2016).

2.1.1. Natural Rubber

The elastic material which is got from the latex sap of trees is termed NR. NR could be vulcanized and finished into numerous types of rubber products. Numerous types of tropical and sub-tropical trees in the regions of South East Asia, Amazon, and Africa produce the milky fluid latex that is in the kind of latex tubes (Boyce and Arruda, 2001). The rubber molecules that exist in these latex tubes are made up of 8 hydrogens and 5 carbon atoms. A huge number of these rubber molecules are merged to form an extensive, chain-like structure. This chain of rubber molecules is termed polymers that provide rubber its property of elasticity (Haupt and Sedlan, 2001; Chester and Anand, 2010).

2.1.2. Synthetic Rubber

Any type of artificial elastomer (a polymer) is termed synthetic rubber. An elastomer could be defined as a material having the property of elasticity. Therefore, the type of rubber formed from chemicals to work as the alternate for NR is a synthetic rubber. There are numerous kinds of polymers utilized for making synthetic rubber kinds. Because of this, different kinds of synthetic rubbers had different properties that are made for the precise needs of rubber products industries (Bokobza and Rapoport, 2002; Teh et al., 2004).

Elastomers have been categorized into groups according to the similarity of applications and properties. Rubber types that had been standardized (SIS 162602, ASTM D 2000, SFS 3551,) are appropriate for numerous industrial applications (for example, tubes, belts, tires, and seals). Certain of the elastomeric groups are described below (Barlow, 1978; Van Beilen and Poirier, 2007):

2.1.3. Rubber Type 61 (Rubbers for General Use)

Type 61 rubbers are utilized when the product doesn't need special properties, like heat oil, or weather resistance. These rubbers have good processability and mechanical properties. They also have less price. Elastomers that fit into this group are NR, IR (polyisoprene rubber), and SBR (styrene-butadiene rubbers) and the blends of these elastomers (Baker et al., 1985; Mooibroek and Cornish, 2000).

2.1.4. Rubber Type 62

Rubber type 62 is a rubber kind that has not been standardized. Butyl rubber (IIR), bromobutyl rubbers (BIIR), and chlorobutyl rubbers (CIIR) are elastomers that fit into this group. They have good weather resistance and ozone. Also, the gas penetrability is low and they are unaffected to vegetable oils, however not to mineral oils (Delzell et al., 1996; Bode et al., 2001).

2.1.5. Rubber Type 63

Rubbers in this group have good oil resistance, however, their weather resistance and ozone are weak. Applications are products that came in connection with oils. NBR (Nitrile rubber) is of rubber type 63 (Mars and Fatemi, 2002; Findik et al., 2004; Zhang et al., 2010).

Rubber type 631 is a rubber that has formed from nitrile rubber. It has better ozone, heat, and weather resistance than nitrile rubber. Hydrogenated nitrile rubber (HNBR) belongs to this group. Rubber type 632 is nitrile rubber merged with polyvinylchloride (NBR/PVC). It has better ozone, oil, and weather resistance than NBR (Polmanteer, 1988; Amin et al., 2006; Polgar et al., 2015).

2.1.6. Rubber Type 64

CR is representative of rubber type 64. It has a good confrontation with vegetable oils and impartially resistance to good naphthenic oils and aliphatic. A drawback is its poor resistance to aromatic oil (Von Hintzenstern et al., 1991; Choi, 2002).

2.1.7. Rubber Type 65

Rubbers in this group have good heat and weather resistance and fairly well oil resistance. ACM (Polyacrylic rubbers) are in this group (Bahia and Davies, 1994; Oshinski et al., 1996).

2.1.8. Rubber Type 66

Rubber type 66 is not uniform. Polyurethane rubbers (EU, AU) belong to this group. These rubbers are hard and have good oil and weather resistance. Their heat resistance is poor (Xiao et al., 2009; Chen et al., 2019).

2.1.9. Rubber Type 67

Rubbers in this group FPM (fluorocarbon rubbers) had good heat, oil, weather, and chemical resistance (Wu et al., 2005; Manosalvas-Paredes et al., 2016).

2.1.10. Rubber Type 68

Q (Silicone rubbers) belong to this group. They had good weather, heat, and cold resistance. Their mechanical properties are feeble (Huang et al., 2004).

2.1.11. Rubber Type 69

Epichlorohydrin rubbers (GECO ECO, CO) belong to this group. They had medium oil, heat, and weather resistance (Von Hintzenstern et al., 1991; Shu and Huang, 2014).

2.1.12. Rubber Type 70

Rubber type 70 includes ethylene-propylene rubbers (EPM, EPDM). They have good ozone, heat, and weather resistance, and weak oil resistance (Horgan and Saccomandi, 2002; Dondi et al., 2014).

2.1.13. Other Rubbers

These rubbers are not standardized (Chang et al., 1989; Qi et al., 2003):

- **Chlorosulfonated Polyethylene (CSM):** Acid resistance and good weather.
- **Ethylene-Vinyl Acetate Copolymer (EVA):** Resistant to aliphatic oils.
- **Butadiene Rubber (BR):** Good elasticity.
- **Chlorinated Polyethylene (CM):** Medium heat and weather resistance.
- **Carboxylated Nitrile Butadiene Rubber (XNBR):** Hard and oil resistant.

2.2. NATURAL RUBBER (NR)

NR could be quarantined from more than 200 diverse species of plants. A commercially important source of NR is Hevea Brasiliensis. NR is gained from latex, which is the blend of 4-polyisoprene, cis-1, and water. Latex is attained from the tree by tapping the inner bark and gathering the latex in cups. A stabilizing agent, like ammoniac, could prevent too early coagulation (Figures 2.1 and 2.2) (Wang et al., 2002).

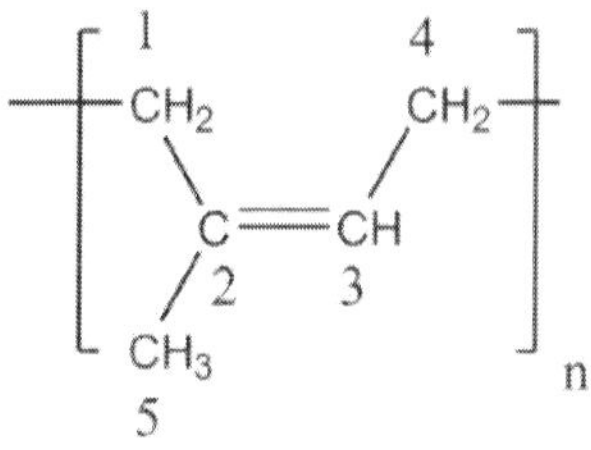

cis-1,4-polyisoprene

Figure 2.1. Two-dimensional chemical structure of4-polyisoprene, cis-1.

Source: https://www.researchgate.net/figure/Left-two-dimensional-chemical-structure-of-cis-1-4-polyisoprene-Right_fig1_320449014.

Figure 2.2. Gathering of the latex from the rubber tree.

Source: https://www.britannica.com/science/rubber-chemical-compound/Tapping-and-coagulation.

Latex could be concentrated through creaming or centrifuging and vended as concentrated latex. Latex could be coagulated with acetic acid or hydrogen carboxylic acid, granulated or formed in sheets, and then dried up to solid raw rubber. Raw rubber kinds are for Instance air-dried sheets (ADS), ribbed smoked sheets (RSS), and pale crepes (Andersen, 1963).

The NR also comprises a few percent of non-rubber ingredients like sugars, fatty acids, resins, proteins, which could function as accelerators and weak antioxidants in the NR. NR is normally vulcanized utilizing sulfur, however, also isocyanates and peroxides could be used (Poh and Ng, 1998).

The biggest manufacturer countries of NR are Indonesia, Malaysia, Thailand, India. Some classification structures that describe the maximum content of cinder, dirt, nitrogen, and volatile elements had been developed in these countries. One renowned system is the Standard Malaysian Rubber system, which had been utilized since 1965.

There are various techniques for processing latex into commercial ranks of latex and dry NR, as revealed in the diagram (Figure 2.3) (Rubber Engineering, Indian Rubber Institute, McGraw-Hill, 2000).

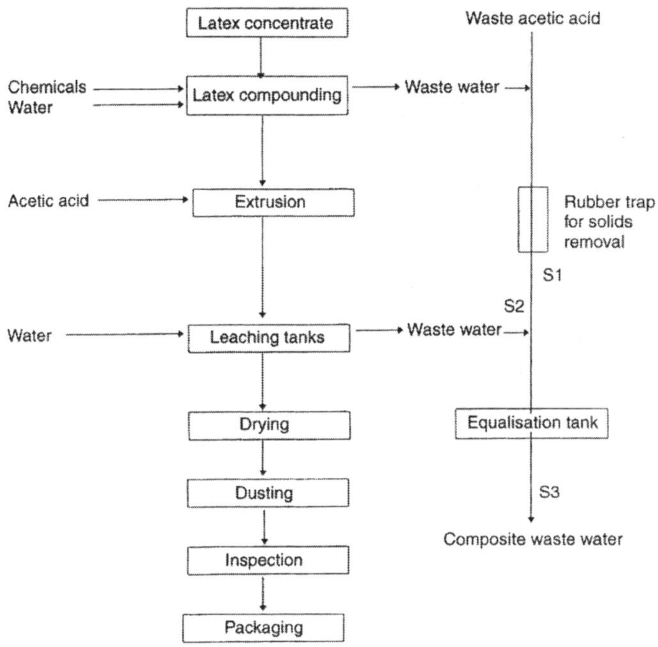

Figure 2.3. Techniques of processing latex into commercial grades of the dry natural rubber.

Source: https://www.researchgate.net/figure/Schematic-ow-diagram-of-the-rubber-thread-manufacturing-process-showing-locations-of_fig1_226197209.

The functioning temperature range for NR is –55 – +70°C.

Advantages of NR are (Laura et al., 2003; Zhao et al., 2016):

- High elongation;
- Excellent cold resistance;
- Good processability;
- Good wear resistance;
- Good tear resistance;
- Good tensile strength;
- Excellent elastic properties;
- Good electrical insulator;
- Little dissipation factor-low heat created in dynamic stress;
- High resistance to acids and water and acids (not to oxidizing acids).

Disadvantages of NR are:

- Swelling in fuels and oils: fuel resistance and low oil;
- Limited high-temperature resistance (short-time maximum temperature 100°C);
- Poor ozone and weather resistance;
- Inappropriate for usage with organic liquids in common (even though vulcanization significantly increases swelling resistance), the main exception being low molecular weight alcohols.

Applications of NR are:

- Footwear;
- Gaskets;
- Tires (60 to 70%);
- Adhesives;
- Tubes, conveyor belts, and v-belts;
- Latex products;
- Coatings.

There are also numerous modified kinds of NR. There is for example. oil-extended natural rubber (OENR) which comprises 20–30% oil, methacrylate grafted NR (Heveaplus MG), and epoxidized NR. These modifications aim to increase the properties of NR to meet the special requirements of rubber manufacturers (Malas et al., 2014).

2.3. ISOPRENE RUBBER, POLYISOPRENE (IR)

Polyisoprene rubber had a similar basic chemical formula to NR, and therefore it is a synthetic version of NR. The study of materials similar to NR was initiated at the beginning of the 20th century, however, because of the weak quality of polymers and the high price of raw materials, industrial production was not activated. Major production was initiated in the 1970s, when economical monomers and catalysts, which create stereo-specific polymers in the solution polymerization became accessible (Sadhu and Bhowmick, 2004).

It is probable to create different types of isomeric structures utilizing different polymerization and catalyst conditions in the polymerization of isoprene monomers. The usable system is cis-1,4-, 3,4-, and trans-1,4-polyisoprenes. Cis1,4-polyisoprene is a synthetic alternate for NR and trans-

1,4-polyisoprene is a tough thermoplastic material (Balata or Gutta-Percha) (Figure 2.4).

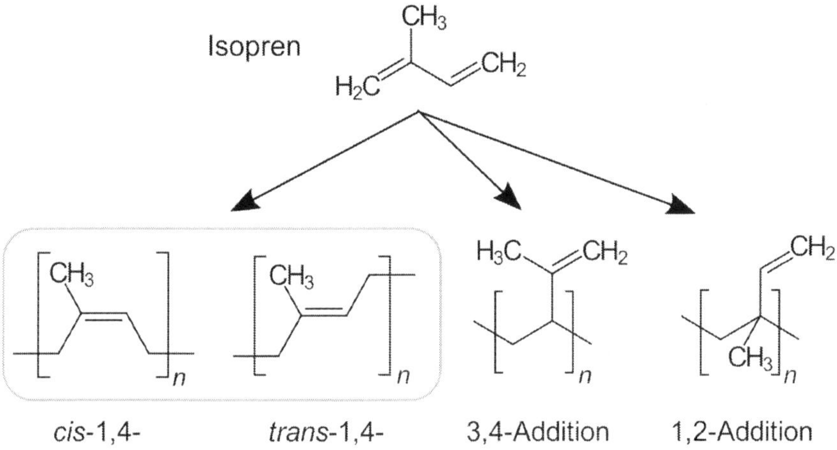

Figure 2.4. The isomeric structures of polyisoprene.

Source: https://en.wikipedia.org/wiki/Polyisoprene.

The properties of polyisoprene rely on the number of its cis-1,4-units. Commercial synthetic isoprene rubbers could be segregated into different groups according to the catalyst utilized (Yoon et al., 2020):

- The Li-IR group, whose catalyst in the polymerization is alkyl lithium. The quantity of cis-1,4-units in Li-IR is around 90% (10% 1,2-type IR). The TiIR group. In these polymerizations, the catalysts are diverse types of Ziegler-Natta catalysts. The usual content of cis-1,4-cis-units in Ti-IR is at level 96% to 98%.
- Lanthanide polyisoprenes had been developed in current years. They estimate very well to NR. The share of 1,4-cis-units in lanthanide IR could be 99.5%.

The quantity of cis-1,4-units affects crystallization and consistency of the molecule structure. With a rise in cis-1,4-content, crystallization is assisted, the glass transition temperature reduces and strength properties enhance. Thus, strength properties like tensile strength, tear-resistance, and modulus, are slightly inferior in synthetic polyisoprenes than in NR, whose cis-1,4-content is nearly 100%. Similarly, the building tack of IR is slightly inferior to that of NR, and the green strength is inferior. Else, the properties of synthetic isoprene rubbers are the same as those of NR. The most important advantages of synthetic polyisoprenes associated with NR are their good processibility, homogeneity, and purity of polymer structure.

Advantages of synthetic IR are:

- Competitive price;
- High tensile strength;
- Good uncured tack;
- Good hot tear strength;
- Cold resistance;
- Toughness;
- High resilience;
- Processability and adherence;
- Good abrasion resistance;
- Resistance to many inorganic chemicals.

Disadvantages of IR are:

- Needs protection against ozone, O_2, and light;
- Non-resistant to hydrocarbons;
- Inappropriate for usage with organic liquids;
- Limited lifetime at high temperatures and in oxidative situations;
- Poor oil resistance.

IR is frequently used with additional rubbers. Through mixing other rubbers with isoprene, tear, and tensile strength and flexibility are increased. Applications of IR are the same to NR which comprises transmission straps and tires conveyor lines, gaskets, paddings footwear, tubes, sports equipment protective gloves, sealing materials, and sealants.

Trans-1,4-polyisoprene (gutta-percha) looks like plastic and is utilized, for example. in deep-sea cables, golf balls, orthopedic applications, and adhesives. Gutta-percha could also be got from the pruning of special trees that are innate to Malaysia.

2.4. BUTADIENE RUBBER (BR), POLYBUTADIENE

The indication of polybutadiene rubbers was Buna, which was made for the first time in 1920 by Germany. Buna was a compound of sodium and butadiene. During World War I, it was observed that the cold resistance of Buna was not sufficient good. Due to this reason, American rubber scientists polymerized BR (polybutadiene) in 1954. BR rubbers had much-enhanced weather resistance than Buna (Noriman et al., 2010).

Utilizing solution polymerization in hydrocarbon solvent usually executes the polymerization of BR. Appropriate catalysts are Ziegler-Natta mixtures and lithium and its compounds. The elastomer is regularly named according to the metal in it or according to its catalyst. Abbreviations utilized are among others Co-BR (cobalt), Ni-BR (nickel), and Li-BR (lithium).

Three different types of basic construction units could be made in polymerization. The polymerization conditions and catalyst disturb the development of these units (Figure 2.5).

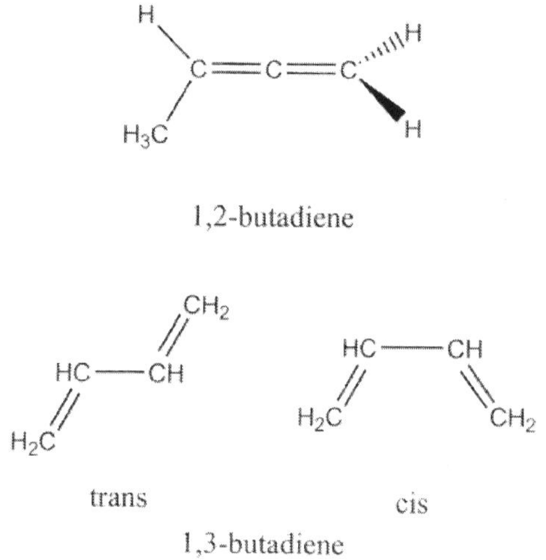

Figure 2.5. The isomeric structures of BR.

Source: https://iopscience.iop.org/article/10.1086/319009/fulltext/52074.fg1.html.

The properties of the polymer are described through the isomeric structures which seem most. BRs could be separated into three groups according to the number of cis-units (Tables 2.1 and 2.2).

Table 2.1. Polybutadiene Rubbers According to the Catalyst Utilized

Characteristics	Li-BR	Ti-BR	Co-BR
Cis-1,4-content (%)	38	93	96
Trans-1,4-content (%)	52	3	3
Glass transition temperature (T_g, °C)	–93	–106	–108
Molar mass distribution	Very thin	Thin	Medium board
Branching degree	Very low	Low	Medium
Melting temperature (T_m, °C)	Amorphous	–22	–11
1,2-content (%)	10	4	1

Table 2.2. Comparison the Properties of NR and BR

Properties	Natural Rubber NR	Butadiene Rubber BR
Hardness (°IRH)	30–90	40–80
Elongation at break (%)	100–700	100–600
Tensile strength at break (N/mm^2)	7–28	7–21
Operating Temperature Range:		
minimum (°C)	–55	–70
maximum (°C)	80	80
Elasticity	5	5
Electrical properties	4	4
Resistance:		
weather and ozone	1–2	1–2
radiation	2–3	2–3
water	5	3
alkalis	2–3	2–3
flame	1	1
abrasion	4–5	4–5
acids	2–3	2–3
Gas permeability	3	3
Tack to the metal	5	5
Adherence	4	4
Residual compression, % (°C)	20–60 (70)	-

Note: 1 = poor, 2 = fair, 3 = good, 4 = very good, 5 = excellent.

Polybutadiene rubbers could be vulcanized with peroxides, sulfur compounds, and sulfur. The peroxide vulcanization is very efficient and generates highly crosslinked polybutadiene rubbers.

Advantages of BR are:

- Abrasion resistance;
- Elasticity;
- Excellent heat resistance and cold resistance;
- Excellent low-temperature resilience and flexibility.

Disadvantages of BR are:

- Weak mechanical properties;
- Poor processability.

The processing of BR is difficult. Therefore, it is generally blended with certain other rubbers, like SBR and NR. In those blends, BR aims to decrease heat build-up and increase the abrasion resistance of the blend. It also increases flexibility.

Applications of BR are:

- Transmission belts;
- Gaskets, tubes;
- Conveyor belts;
- Shoe soles;
- Tires (BR content usually 30 to 50%, blended with NR and SBR);
- Coatings;
- Coatings of cylinders, V belts;
- Toys.

2.5. STYRENE-BUTADIENE RUBBER (SBR)

SBR (Styrene-butadiene rubber) is the most significant sort of synthetic rubber. It was primarily developed to substitute NR. The manufacturing technique of SBR copolymer was established in 1929 in Germany when the emulsion polymerization technique at around 50°C became the master. In that technique, the macromolecular amorphous copolymer is polymerized with butadiene and styrene (Zhang et al., 2005; Noriman et al., 2010).

There exist four diverse basic construction units in the SBR, and three of them initiate from butadiene (Figure 2.6).

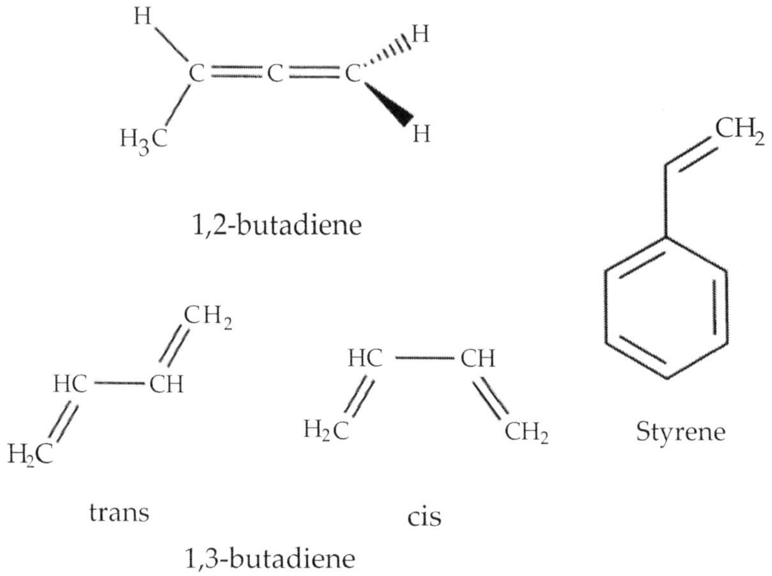

Figure 2.6. The structure of styrene and isomeric structures of polybutadiene.

Source: https://iopscience.iop.org/article/10.1086/319009/fulltext/52074.fg1.html.

Currently, styrene-butadiene elastomers could be produced through emulsion or solution polymerization methods. "Cold" emulsion polymerization, at around 5°C, is the most extensively utilized polymerization method, even though the solution technique had steadily enhanced its market share.

In solution polymerization, the polymerization usually happens in dry hydrocarbon solvents along with anionic methyl-lithium catalyst. Relying on the specific polymerization procedure, two different elastomer kinds could be formed. One type comprises TPE (butadiene blocks) and segmented styrene, and the other kind is rubber elastomer with an unusual distribution of co-monomers in a polymer (Mousa and Karger-Kocsis, 2001).

SBR could be vulcanized utilizing peroxides, sulfur donor systems, and sulfur. The processing technique could affect the properties of SBR significantly. Styrene content, molar mass, and the number of units differ, relying on the manufacturing method. Instances are shown in Table 2.3.

Table 2.3. Properties of Solvent Polymerization and Emulsion Polymerization

Properties	***Solvent-SBR***	***Emulsion-SBR***
Molar mass (M_n, g/mol)	*200000*	*145000*
Trans-1,4-content (%)	*54*	*65*
M_w/M_n	*2.1*	*4.5*
Cis-1,4-content (%)	*35*	*18*
Styrene content (%)	*18*	*23.5*
1,2-content (%)	*11*	*17*
Glass transition temperature (T_g, °C)	*-69.7*	*-50.6*
Molar mass (M_w, g/mol)	*420000*	*651000*

The commercial products of SBR are Kaylene, Cariflex, and Buna EM.

The type designation as per the numeric code (Table 2.4):

- 10xx hot polymer without filler;
- 12xx solution-SBR;
- 15xx cold polymer without filler;
- 16xx cold polymer, carbon black masterbatch;
- 17xx hot polymer, oil-extended;
- 18xx cold polymer, oil masterbatch/carbon black;
- 19xx emulsion resin masterbatch;
- 'xx' specifies coagulant, viscosity, content of styrene.

Advantages of SBR rubbers are:

- Low price;
- Good elasticity;
- Good abrasion and aging resistance.

Disadvantages are:

- Poor ozone resistance;
- Poor oil resistance;
- Inferior mechanical properties (needs reinforcements);
- Low elongation at break;
- Don't resist aromatic, halogenated, or aliphatic solvents;
- Adhesion properties.

Table 2.4. Comparison Amongst the Properties of NR and SBR

Properties	*Natural Rubber (NR)*	*Styrene-butadiene-Rubber (SBR)*
Hardness (°IRH)	*30–90*	*40-.90*
Elongation at break (%)	*100–700*	*100–600*
Tensile strength at break (N/mm²)	*7–28*	*7-.25*
Operating Temperature Range:		
minimum (°C)	*–55*	*–45*
maximum (°C)	*80*	*100*
Elasticity	*5*	*5*
Resistance:		
radiation	*2–3*	*2–3*
abrasion	*4–5*	*4*
weather and ozone	*1–2*	*1–2*

Note: 1 = poor, 2 = fair, 3 = good, 4 = very good, 5 = excellent.

SBR requires more support than NR to attain tear strength and good tensile and durability. SBR also had lower flexibility than NR.

Applications of SBR are:

- Hoses;
- Sponge and foamed products;
- Car tires (blended with IR, NR, and BR);
- Belting;
- Waterproof materials;
- Footwear;
- Conveyor belts;
- Adhesives;
- Toys;
- Molded rubber goods.

Changing the monomer content in SBR copolymer, utilized in tire tread blends, could amend the properties of tires (Tables 2.5 and 2.6).

Table 2.5. Monomer Contents That are Usual of Tire Tread Blends

The Properties of Polymer	Good Wet Grip	Low Rolling Resistance	General Use
Vinyl content (%)	50	10	35
Styrene content (%)	23	15	20

Table 2.6. The Influence of Monomer Content and Tg:n on the Properties of Tires

The Properties of Tires	Higher Tg	Growing Vinyl Content	Growing Styrene Content
Wet grip	Increasing	Increasing	Increasing
Wet steerability	Increasing	Increasing	Increasing
Fuel consumption	Increasing	Increasing	Increasing
Dry grip	Increasing	Increasing	Increasing
Snow grip	Decrease	Decrease	Decrease
Ice grip	Decrease	Decrease	Decrease
Dry steerability	Increasing	Increasing	Increasing
Lifetime	Decrease	Decrease	Decrease

2.6. BUTYL RUBBERS

Butyl rubbers comprise CIIR, BIIR, and Isobutylene-Isoprene Rubber (IIR). Butyl rubbers are made by copolymerizing little amounts of isoprene with isobutylene. Isoprene units are positioned casually in the isobutylene chain in trans-1,4 form. Regulating the polymerization temperature and the proportion of monomers could differ the composition of the polymer. A usual butyl rubber includes 0.5–3 mole percent isoprene (Cheng et al., 1990).

The properties of butyl rubbers rely on the saturation degree and the length of the molecule chains. When the quantity of double bonds is low, rubber had ozone resistance and good oxygen. A greater quantity of double bonds increases the vulcanization procedure and also enhances the number of crosslinks (Figure 2.7) (Van Dyke et al., 2003).

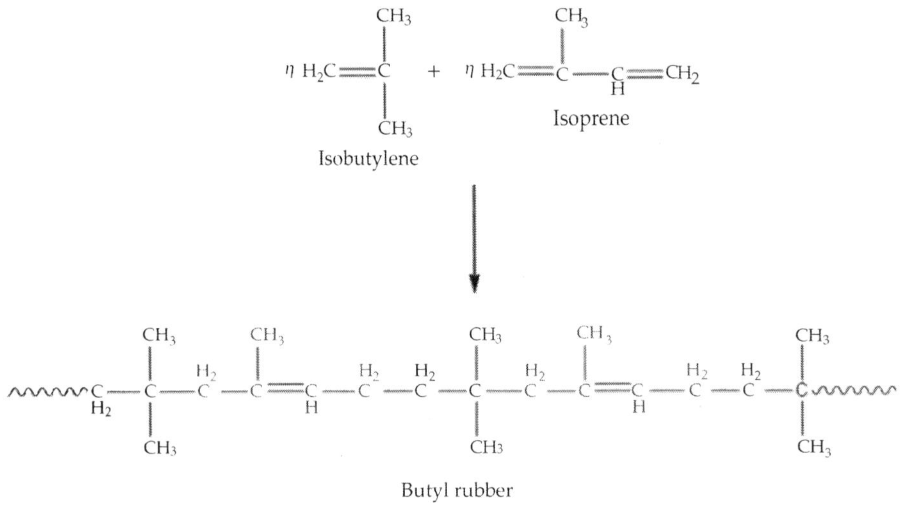

Figure 2.7. Butyl rubber synthesized from isoprene units and isobutylene.

Source: https://www.chegg.com/homework-help/isobutylene-isoprene-co-polymerize-give-butyl-rubber-draw-str-chapter-26-problem-12p-solution-9780321971128-exc.

The properties of butyl rubbers could be enhanced by adding 1 to 2 weight percent of halogens and by creating bromobutyl (BIIR) and chlorobutyl (CIIR) rubbers. Halogens are typically combined to the double-bonded carbon deprived of the methyl group in the isoprene unit. The adding of the halogens enhances chain flexibility and improves cure compatibility in blends with other diene rubbers.

The butyl rubber could be cured with sulfur, however, it required an accelerator. Dioxide compounds collected with an oxidizing agent could also be utilized. In that situation, crosslinks stand heat superior to sulfur bonds. BIIR and CIIR had more reactive.

Points in the crosslinking if the curing agent (metal oxides or sulfur) had been utilized in curing. Peroxides can't be used, because they might discontinue the elastomer chains.

Advantages of butyl rubbers are:

- Good weather resistance;
- Low water absorption;
- Heat stability;
- Stabile in high temperatures and at long-term-use;
- Good ozone resistance;
- Resistant to oxidizing agents, animal fats, and vegetable and polar solvents;
- Low gas permeability;
- Elasticity in the wide temperature range –73–100°C.

Disadvantages of BR are:

Relatively low elasticity;

Poor wear resistance;

Not resistant to hydrocarbon solvent and oil.

The properties of halogenated butyl rubbers are the same as those of basic butyl rubber. Though, they had lower gas permeability and enhanced ozone, thermal, chemical, and weather resistance. Halogenated butyl rubbers are utilized in applications that need rubber with a high vulcanization rate.

Applications of BR are:

- Waterproof films;
- Inner tires of bicycles and cars;
- Inner tubes;
- Steam hoses;
- Coatings of cables and fabrics;
- Vibration isolation;
- Base element of chewing gum;
- Pharmaceutical closures and membranes;
- Gutter gasket.

2.7. NITRILE RUBBER, NITRILE-BUTADIENE RUBBER, AND ACRYLONITRILE RUBBER (NBR)

Poly-acrylonitrile-butadiene rubber is a copolymer of acrylonitrile and butadiene. It was produced for the first time in 1930. It is utilized because of its good fuel, oil, and fat resistance. Acrylonitrile rubbers are also termed just nitrile rubber. NBR is formed by emulsion polymerization. The polymerization rates of butadiene and acrylonitrile are different. Due to that, the content of monomers in the reaction mixture is not the same as the content of monomers in the copolymer. The polymer made is a random copolymer in which the acrylonitrile content fluctuates between 18 and 50%. Feeding with monomers or changing the temperature could modify the composition of the polymer (Figure 2.8) (Liu et al., 2016).

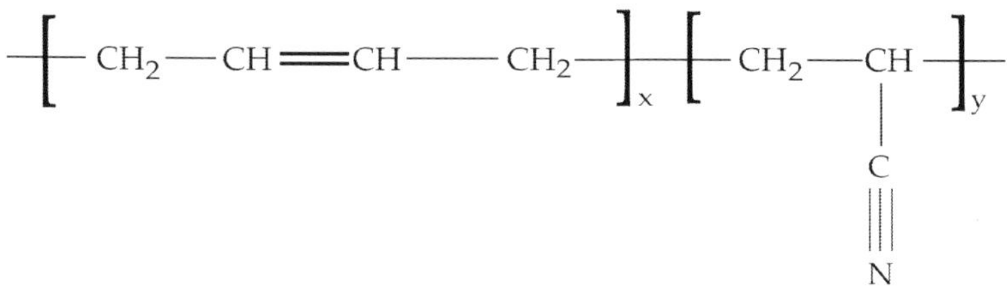

Figure 2.8. Acrylonitrile and butadiene units.

Source: https://laroverket.com/wp-content/uploads/2015/03/Elastomeric_materials.pdf.

Enhancing the acrylonitrile content improves hardness, abrasion resistance, oil resistance, and heat resistance, however, increases the glass transition temperature.

Despite most other synthetic rubbers, nitrile rubbers could be vulcanized with numerous crosslinking systems. The vulcanization could take place at high temperatures or at room temperature to accelerate the reactions (Zhu et al., 2005).

Nitrile rubbers are utilized in applications that claim good mechanical properties and fuel and oil resistance. NBR could be used mixed with other rubbers. For example, the enhancing of IIR to NBR increases thermal stability and weather properties and reduces the gas permeability of NBR.

Properties of NBR are:

- Low ozone resistance;
- Good resistance to aliphatic, oil, and aromatic hydrocarbons and vegetable oils;
- Good abrasion and water resistance;
- High oil and heat resistance;
- High swelling with certain solvents (esters and ketones) and some oils.

Applications of NBR are:

- Conveyor belts;
- Containers;
- Hoses, joints, seals;
- Protective shoes and clothes;
- Roll coverings.

2.7.1. Modified Nitrile Rubbers

2.7.1.1. HNBR and XNBR

Hydrogenated nitrile rubbers (HNBR) and Carboxylated nitrile rubbers (XNBR) are special amendments of NBR. The XNBR rubbers comprise randomly positioned carboxyl groups that are derived from acrylate acid or methacrylate acid. The XNBR had better hardness, tensile strength, and abrasion resistance. It also had improved low-temperature brittleness and enhanced retention of physical properties after air aging and hot-oil compared to NBR (Bieliński et al., 1998; Vijayabaskar et al., 2006).

2.7.1.2. Hydrogenated Nitrile Butadiene Rubber (HNBR)

The nitrile rubber could also be improved through (partially) saturating the double bonds in major chain butadiene through catalytic hydrogenation. This type of HNBR, NBR, had been developed to repel better aging in hot air and oil. Key applications of nitrile rubber comprise seals, cables, vehicle tubing, and profiles.

Properties of HNBR are:

- Application temperature up to 150°C;
- High tensile strength, weather-resistant;
- Gasoline and oil swelling as for NBR;
- Peroxide curable kinds (double bond content < 1%) and sulfur curable (double bond content < 4–6%).

2.8. EPICHLOROHYDRIN RUBBERS

Epichlorohydrin rubbers comprise epichlorohydrin/ethylene oxide copolymer (ECO), CO (epichlorohydrin homopolymer), and epichlorohydrin terpolymer (ETER). There are three different kinds of epichlorohydrin elastomers: CO (epichlorohydrin homopolymer), ECO (epichlorohydrin/ethylene oxide copolymer), and epichlorohydrin terpolymer (ETER), which form from ethylene oxide, epichlorohydrin, and certain other monomers (typically diene) (Figure 2.9).

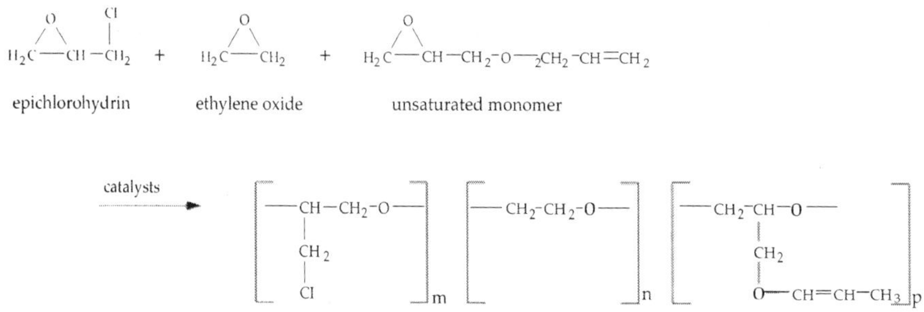

Figure 2.9. The polymer made from the structures of ECO and CO.

Source: https://www.sciencedirect.com/topics/chemistry/epichlorohydrin.

In the polymerization of epichlorohydrin, a coordinate catalyst is utilized. The catalyst could be for instance a compound of water, acetylacetone, and aluminum alkyl. The polymerization technique utilized is solution polymerization in the hydrocarbon solution. When vulcanizing copolymer and homo-, chloromethyl groups react with a difunctional curing agent, like urea or ethylene thiourea, diamine. Terpolymers could be vulcanized with peroxide or sulfur (Panigrahi et al., 2019).

The main differences between copolymer (CO) and epichlorohydrin homopolymer (ECO) are in cold resistance and elasticity. ECO is very elastic over an extensive temperature range, while CO is elastic only at high temperatures. That is why the epichlorohydrin copolymers are utilized more than homopolymers.

Properties of epichlorohydrin rubbers are given below:

- Good processability;
- High price;
- Low gas permeability;
- High cold and heat resistance;
- Can cause corrosion with metal;
- Resistance to fuels, oils, and chemicals;
- Good damping properties;

- Good fire resistance;
- Good weather, ozone, and thermal resistance;
- Weak tensile strength (fillers reinforce);
- Very good dynamic properties.

The usage of epichlorohydrin rubbers is the same as that of nitrile rubbers. Though, ECO gives better elasticity, processability, and oil resistance.

The applications of epichlorohydrin rubbers are:

- Coatings of textiles;
- Vibration isolator;
- Gaskets;
- Membranes;
- Petrol and oil tanks and hoses;
- Resilient mountings;
- Belts, rolls;
- Coatings of cables and wires.

2.9. ETHYLENE-PROPYLENE RUBBERS

Ethylene-propylene rubbers could be separated into two groups: EPDM (ethylene-propylene-diene rubber) and EPM. EPM is a copolymer of propylene and ethylene and EPDM is a terpolymer of diene, ethylene, propylene. The most commonly utilized dienes which provide the crosslinking sites for the elastomer are ethyldienenorborne, dicyclopentadiene, and 1,4-hexadiene (see formulas below).

Rubbers normally contain 45 to 60 wt.% of ethylene monomer. Material with low ethylene contented is easier to proceed with than high ethylene contented material. Especially extrudability and green strength enhanced as the ethylene content increases. Diene content is generally 4–5%; however, sometimes it could be even 10% (Figure 2.10).

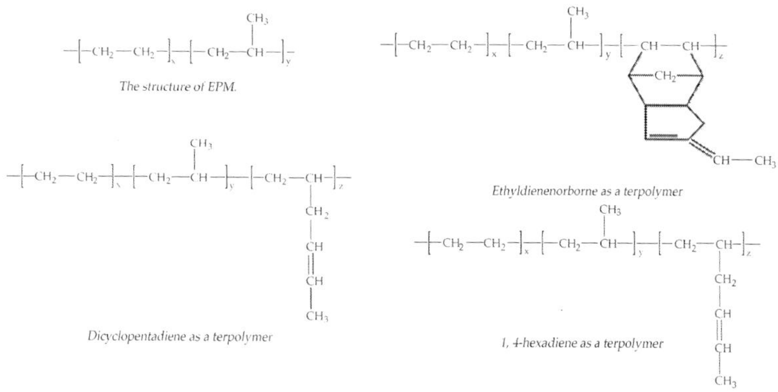

Figure 2.10. The structure of EPM rubbers.

Source: https://laroverket.com/wp-content/uploads/2015/03/Elastomeric_materials.pdf.

Ethylene-propylene rubbers are formed mostly through solution polymerization with Ziegler-Natta type catalysts. EPM rubbers can't be vulcanized with sulfur due to the lack of unsaturation in the main chain. EPM could be cured with radiation or peroxides. EPDM could be vulcanized with peroxide, sulfur, radiation, and resin cures. Catalyst technologies and polymerization in use currently provide the capability to design polymers to fulfill demanding and specific processing and application needs (Aminabhavi and Khinnavar, 1993).

2.9.1. Typical Properties

Ethylene-propylene rubbers are valued for their outstanding resistance to heat and their oxidation, weathering, and ozone resistance because of their stable, saturated polymer backbone structure. Non-black compounds and properly pigmented black are color-stable.

As non-polar elastomers, they had good electrical resistivity also as resistance to polar solvents like alkalis, phosphate esters, acids, water, acids, and many alcohols and ketones.

Low crystalline grades or amorphous had excellent low-temperature elasticity with glass transition points of about –60°C (Danesi and Porter, 1978).

Heat aging resistance up to 130°C could be attained with properly nominated sulfur acceleration structures and heat resistance at 160°C could be attained with peroxide cured compounds. Compression set resistance is worthy, specifically at high temperatures, if peroxide cure systems or sulfur donors are utilized. These polymers reply well to plasticizer loading and high filler, providing cost-effective (obs. low density too), easily processible compounds. They could have tear properties and develop high tensile, excellent abrasion resistance, also as enhanced flame retardance (Phiri et al., 2015). As the drawbacks of EP rubbers, hydrocarbon resistance and bad oil, and poor tack could be mentioned. A broad summary of properties (property ranges) is presented in Table 2.7.

Table 2.7. Properties of Ethylene-Propylene Rubbers*

Property	***Value Range***
Ethylene content (wt.%)	*45 to 80 wt.%*
Diene content (wt.%)	*0 to 15 wt.%*
Elongation (%)	*100 to 600*
Abrasion resistance	*Good to excellent*
Hardness, Shore A Durometer	*30A to 95A*
Tensile strength (MPa)	*7 to 21*
Tear resistance	*Fair to Good*
Compression Set B (%)	*20 to 60*
Useful temperature range (°C)	*–50° to +160°*
Electrical properties	*Excellent*
Mooney viscosity, ML 1+4 @ 125°C	*5–200+*
Resilience	*Fair to Good (stable over wide temp. ranges)*
Specific gravity (gm/ml)	*0.855–0.88 (depending on polymer composition)*

*Range could be extended through proper compounding. Not entire of these properties could be attained in one compound.

Applications of ethylene-propylene rubbers are:

- Roll covers;
- Agricultural equipment: seed tubes, hoses, silus, cushioning;
- Wire and cable;
- Hosepipes and gaskets, liners in the building industry;
- Products of the automotive industry: seals and hoses, isolators.

2.10. CHLOROPRENE RUBBER, POLYCHLOROPRENE (CR)

Polychloroprene was one of the initial synthetic rubbers. The first chloroprene monomers were made from acetylene. Currently, they are produced from butadiene, because it is a safer and easier route. Chloroprene is polymerized through emulsion polymerization utilizing potassium (K) persulfate as an opened radical initiator. The key component of the polymer normally is trans-1,4-units. In the vulcanizing of CR, magnesium oxide and zinc oxide blend are usually utilized (Figure 2.11) (Le Gac et al., 2012).

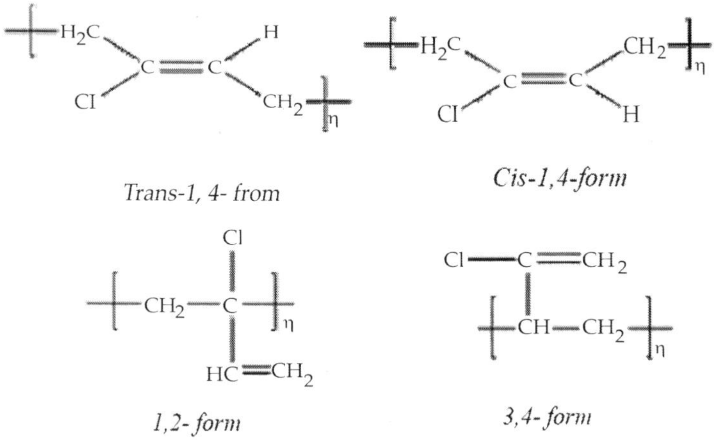

Figure 2.11. Isomeric structures of CR.

Source: https://www.britannica.com/topic/industrial-polymers-468698/Polychloroprene-chloroprene-rubber-CR.

CRs could be separated into W and G types according to their mechanism for regulating the molecular weight of the polymer throughout the polymerization. In G kinds, sulfur is copolymerized with the chloroprene, when it doesn't need acceleration throughout curing. The G-kind rubbers had marginally inferior aging resistance, however, tack, and resilience are better than in the W kinds. The W kinds of CRs need an accelerator. The vulcanization can't be taken out with sulfur. Appropriate accelerators are metal oxides. The W kind rubbers had better thermal resistance and aging properties than G-kind rubbers (Manohar et al., 2017). Polychloroprene is a multipurpose elastomer. It is utilized particularly in demanding circumstances.

The advantages of CR are:

- Inflammability;
- Good tear strength;
- Good abrasion resistance;
- Good adhesion to metals;
- Good ozone resistance;
- Increased hardness in high-temperature environments;
- Good oil and solvent resistance.

The disadvantages of CR are:

- High swelling in hot water, acids, some oils, and some organic solvents.

The applications of CRs are:

- Adhesive;
- Footwear;
- Gaskets;
- V belts and conveyor belts;
- Coated fabrics;
- Hoses;
- Wear suit applications, inflatables;
- Cable and wire coverings;
- Vibration isolators.

2.11. POLYACRYLATE RUBBERS (ACM)

Polyacrylate rubbers are elastomers that are made from acrylic esters (typically methyl and ethyl acrylate) and reactive cure site monomer (chloroethyl vinyl ether or carboxylic acid) (Table 2.8, Figures 2.12 and 2.13).

Table 2.8. The Simple Monomers of Acrylate Rubbers

Monomer	Structure, X
Butyl acrylate	C_4H_9
Ethyl acrylate	C_2H_5
Ethoxy ethyl acrylate	$C_2H_4OC_2H_5$
Methoxy ethyl acrylate	$C_2H_4OCH_3$

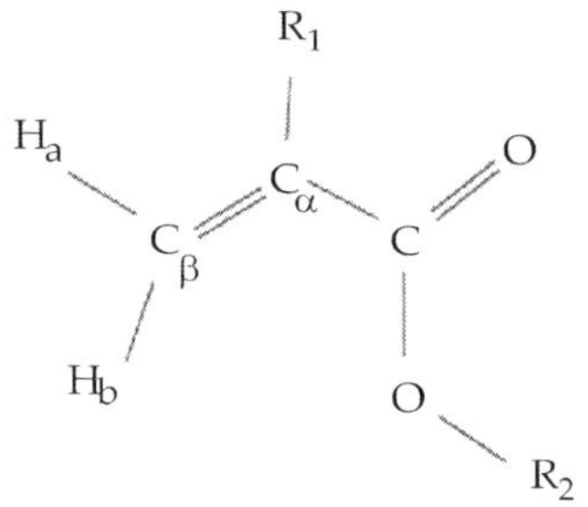

R_1 : H for acrylates; CH_3 for methacrylates

R_2: alcohol moiety

Figure 2.12. The simple structure of acrylates.

Source: https://www.researchgate.net/figure/Structures-of-acrylates-and-meth-acrylates_fig1_40696788.

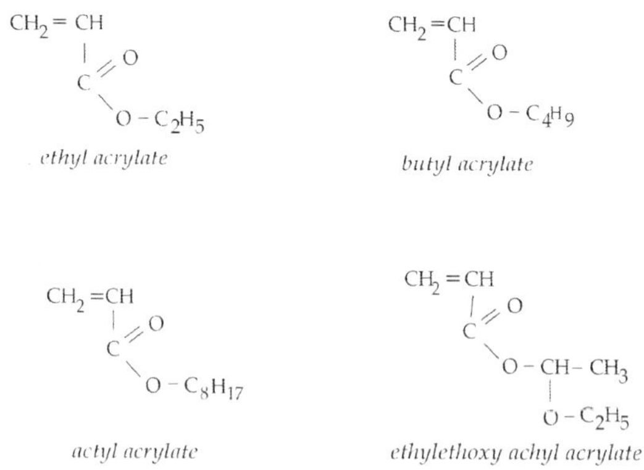

Figure 2.13. Instances of the structure of acrylate rubbers.

Source: http://kompozite.com/2013-09-26/acrylic-rubbers.html.

The preparation of polyacrylate rubbers is founded on the polymerization of methacrylate acids and acrylate. The polymerization method could be precipitation or emulsion polymerization. In emulsion polymerization, the catalyst could be a redox system or persulfate salt. In precipitation polymerization, the catalyst could be peroxide. The peroxides are solvents to atso-bis-isobytyro-nitril or monomer, which reduce easily (Kader and Bhowmick, 2003).

To create reactive sites for the vulcanization, polyacrylate elastomers are copolymerized with 1 to 5 weight percent reactive constituent, like chloroethylene vinyl ether or carboxylic acid, or epoxy compounds. General vulcanization agents are hexamethylene diamine carbamate or metalcarboxyl soaps or methylene dianiline like potassium or sodium-stearate. Sulfur works as a catalyst (Dos Santos and Batalha, 2010).

Properties of ACM are:

- Excellent flexing properties;
- Low gas permeability;
- Good elasticity;
- Low resistance to hot water;
- Excellent ozone and weathering resistance;
- Good heat aging resistance;
- Very good heat resistance;
- Good oil resistance;
- Not highly corrosive to steel;
- Resistant to oil and aliphatic solvents;
- Poor alkali, water, and acid resistance.

Applications of ACM are:

- Hoses, seals, wire coverings;
- Applications in the automotive industry (for example, grommets, Boots, and seals);
- Adhesive formulations.

2.12. POLYURETHANE RUBBERS (AU, EU, PUR)

Polyurethanes (PU) are termed after the urethane group, which creates when the isocyanate group reacts with the hydroxyl group of the alcohol.

Relying on the amount and type of feeding stocks and additives, PUs could be thermoplastics or thermosets (Figure 2.14) (Yamasaki, 2016).

Figure 2.14. Creation of urethane group.

Source: http://kompozite.com/2013-09-26/acrylic-rubbers.html.

PUs are the only most flexible family of polymers there is. PUs could be microcellular elastomers or solid (both TPEs and crosslinked rubbers), paints, foams, paints, adhesives, or fibers. They could also be processed with the greatest processing techniques known at present (Mahmood et al., 2012).

Urethane thermoplastic elastomers (TPE-U) and also polyurethane rubbers (PUR) are made up of long, short, and soft segments, hard segments. The soft segments are made through the reactions between polyether diol or polyester diol with hydroxyl group ends. The hard segments are made through the reactions between chain extenders and isocyanates. The PUR could be separated into polyether urethane rubber (EU) and polyesterurethane rubber (AU) according to the polyol utilized (Figures 2.15 and 2.16).

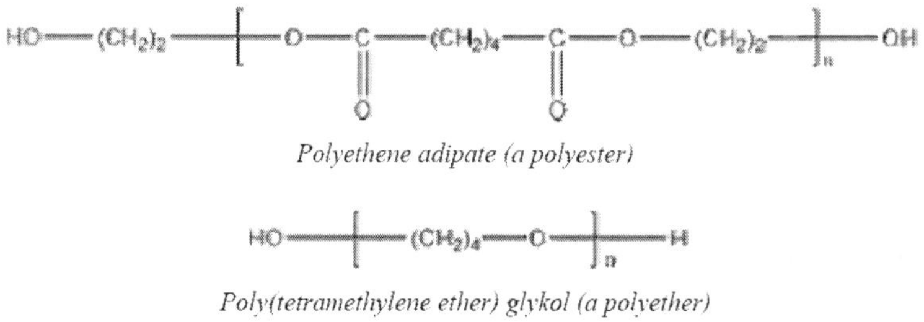

Figure 2.15. Typical polyols utilized in polyurethanes.

Source: http://kompozite.com/2013-09-26/acrylic-rubbers.html.

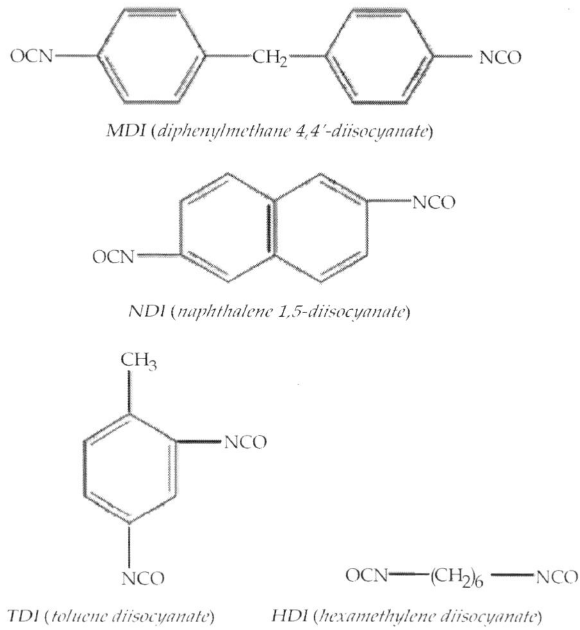

Figure 2.16. Typical diisocyanates that are utilized in polyurethanes.

Source: https://docplayer.net/36571488-Elastomeric-materials.html.

The PUR could be separated into kneaded (millable) and castable PUs according to their processing technique.

Castable PUR are attained in a one-step procedure or a two-step procedure. In the one-step casting technique di-isocyanate, polyol, and chain extender react and the product is made in a similar step. In the two-step casting technique, a prepolymer is formed first by the reaction between polyol and diisocyanate. In the second step, the length of the chains and the molar mass of the prepolymer are enhanced, and the structure is crosslinked with the chain extenders. The second step is frequently taken out in a mold at high temperatures. Extenders might be triols or diols. The two-step casting is additionally utilized than one-step casting (Hu et al., 2008; Lili et al., 2009).

The crosslinking which creates the three-dimensional network in PUR could be brought out, as explained above, through isocyanates or multifunctional chain extenders, however also with peroxides and sulfur (particularly the kneaded PUR grades).

The properties of PUR rely on the structure of their chains. Polyester-based PUR normally had better chemical resistance and mechanical properties than polyether-based PUR. Polyether-based PUR had better hydrolysis resistance and better properties in low temperatures.

Properties of polyurethane rubbers are:

- Hardness;
- Low friction coefficient;
- Good insulator;
- Resistant to oils and aliphatic hydrocarbons;
- Good tensile strength;
- Good abrasion and tear resistance;
- Good oxygen and ozone resistance.

Applications of polyurethane rubbers are:

- Seals;
- Soles;
- Wearing surfaces of rollers and wheels;
- Power transmission elements.

2.13. FLUOROCARBON RUBBERS (FKM, FPM)

Fluorocarbon rubbers are very steady materials due to the strength of the bond between carbon and fluorine. The most usual grades of fluorocarbon rubbers are founded on hexafluoropropylene HFP and vinylidene fluoride monomers (see Table 2.9), which are referred to as FPM in ISO standards and FKM in ASTM standards. There are also fluorocarbon rubbers comprising chlorine in vinylidene monomers (for example, $CFCl = CF_2$), mentioned as CFM rubbers. Fluorocarbon rubbers are generally produced by emulsion radical polymerization. Peroxide compounds work as initiators (Table 2.10 and Figure 2.17) (Li et al., 2017; Yang et al., 2019).

Table 2.9. Monomers Utilized in Fluorocarbon Rubbers

Monomer	**Structure**
Vinylidene fluoride (VF_2)	$H_2C{=}CF_2$
Hexafluoropropylene (HFP)	$F_2C{=}CF$ $\vert$ CF_3
1-hydropentafluoropropylene (HPTFP)	$HFC{=}CF$ $\vert$ CF_3
Chlorotrifluoroethylene (CTFE)	$F_2C{=}CF$ $\vert$ CI
Tetrafluoroethylene (TFE)	$F_2C{=}CF_2$
Perfluoromethylvinylether (FMVE)	CF_2 $\Vert$ $F{-}C{-}O{-}CF_3$

Table 2.10. Structures of Fluorocarbon Rubbers

Monomers	**Structural Unit**	**Type Designation**	**Commercial Types**
VF 2 + HFP	$-CH_2-CF_2-CF_2-CF-$ $\vert$ CF_3	FKM	Viton A, AHV, A-35, E-60, Fluorel 2140, 2141, 2143, 2146, SFF-26
PFMVE + TFE + X	$-CF_2-CF-CF_2-CF_2-X-$ $\vert$ O $\vert$ CF_3	FKM	–
VF 2 + HPFP + TFE	$-CH_2-CF_2-CHF-CF-CF_2-CF_2-$ $\vert$ CF_3	FKM	Tecnoflon T

VF 2 + HFP + TFE	$-CH_2-CF_2-CF_2-CF(CF_3)-CF_2-CF_2-$	FKM	Viton B, B-50
VF 2 + HPFP	$-CH_2-CF_2-CHF-CF(CF_3)-$	FKM	Tecnoflon SL, SH
VF 2 + TFCIE	$-CH_2-CF_2-CFCl-CF_2-$	CFM	KEL-F 3700, 5500, SKF-32

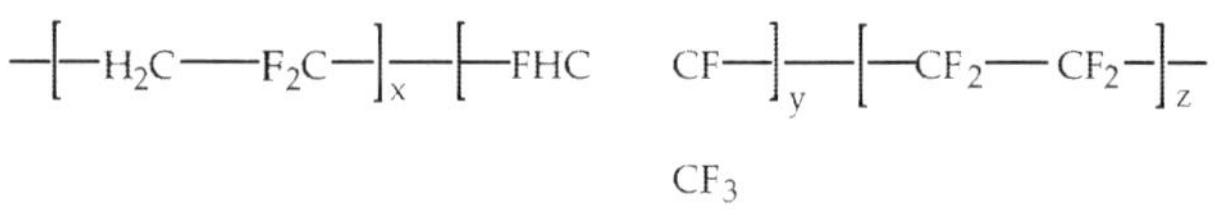

Figure 2.17. An instance of a structure of the fluorocarbon rubbers, VF_2/HPTFP/TFE copolymer.

Source: https://www.dupont.com/content/dam/dupont/amer/us/en/transportation-industrial/public/documents/en/Perfluoroelastomer_and_Fluoroelastomer_Seals_for_Photovoltaic_Cell_Manufacturing_Processes.pdf.

The most generally utilized FKM rubbers could be vulcanized with bisphenols, diamines, and polyhydroxide compounds. The vulcanization system had a metal oxide as an acid acceptor.

Advantages of fluorocarbon rubbers are:

- Incombustible;
- Good abrasion resistance;
- Excellent ozone, oxygen, and weather resistance;
- Excellent heat resistance (temporarily 315°C, up to 200°C);
- Good high-temperature compression-set resistance;
- Good chemical and solvent resistance.

Disadvantages of fluorocarbon rubbers are:

- Limited elasticity at low temperatures;
- High price;
- Relatively poor mechanical properties;
- Low alkali resistance;
- The tensile strength reduces substantially at high temperatures.

The fluorocarbon rubbers are utilized for distinctive applications that need good oxygen heat, or corrosion resistance and oil resistance, and hot solvent.

Applications of fluorocarbon rubbers are:

- Heat-resistant insulators;
- Fire-resistant coverings;
- Airplane and Car seals and hoses;
- Gaskets, valve-stem seals, fuel hoses;
- Shaft seals, O-rings.

2.14. SILICONE RUBBERS (Q)

SRs are inorganic polymers, as their key chain structure doesn't comprise carbon atoms. As revealed in the diagram, silicone, and O_2 atoms-siloxane groups-create the polymer main chain. There are usually also certain pendant groups, typically methyl groups, attached to the polymer chain. The molar mass of SRs could fluctuate over an extensive range, and therefore there are liquid materials also as traditionally resinous rubbers obtainable (Figure 2.18) (Lacy et al., 1981).

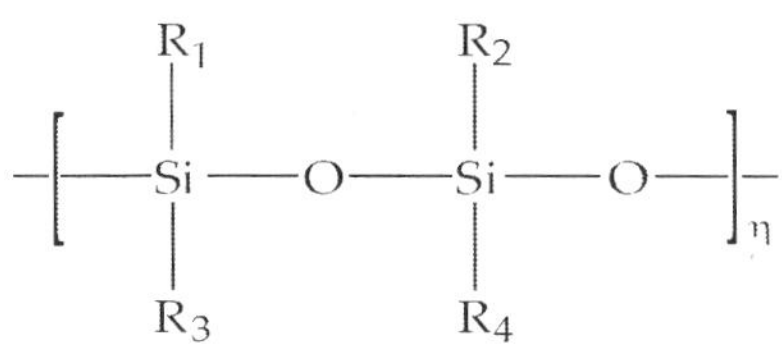

Figure 2.18. The structure of silicone.

Source: https://omnexus.specialchem.com/selection-guide/silicone-rubber-elastomer.

SRs are typically polymerized from cyclic oligomers to linear macromolecules. The vulcanization could be taken out at elevated temperature or room temperature. Vulcanization at room temperature happens with a crosslinking agent (for example, ortho-silicon acid ether) or air. For high temperatures, vulcanization peroxides are utilized (Fukuda et al., 2013). The molar mass of SR vulcanized at high temperatures is higher (300,000 to 1000,000 g/mol) than in room temperature vulcanization (10,000–100,000 g/mol) (Table 2.11).

Table 2.11. Silicone Rubbers and Their Pendant Group Structure

Pendant Group	**Structure**	**Rubber Type**
Methyl	CH_3	MQ
Vinyl	$CH_2 = CH$	VMQ
Phenyl	C_6H_5	PMQ
Trifluoropropyl	$CF_3CH_2CH_2$	FMQ
Vinyl phenyl	$CH_2 = CH\ C_6H_5$	PVMQ
Vinyl trifluoropropyl	$CH_2 = CH\ CF_3CH_2CH_2$	FMVQ

In VMQ rubbers, few of the methyl groups (< 0.5%) are substituted with vinyl groups. This facilitates vulcanization and decreases the deformation set of the rubber. PVMQ and PMQ rubbers had phenyl groups (5 to 10%) rather than methyl groups. This enhances the properties of SRs at low temperatures. Fluorosilicones (FMQ and FMVQ) had enhanced solvent resistance than other SRs (Henderson, 1993).

Reinforcement fillers, like silica, had to be utilized, since the mechanical properties of clean SR are relatively weak. For instance, the tensile strength of unalloyed SR is inferior to that of any other ruccer. Though, the mechanical properties of SR don't weaken at enhanced temperatures as much as in the state of other rubbers.

Advantages of silicone rubbers are (Yang et al., 2017):

- Good release properties;
- Good aging resistance at high temperatures;
- Elasticity;
- High-temperature resistance, extensive operating temperature range (even –100–+300°C);
- Odorless, non-toxic, tasteless;
- UV light, ozone, and O_2 resistance (peroxides have to be utilized for vulcanization);
- Good electrical insulation;
- Good resistance to low concentrations of salts, bases, and acids.

Disadvantages of silicones are:

- Weak mechanical properties without additives;
- Large shrinkage in molded articles;
- High price;
- Vulcanization to get good mechanical properties had to be taken out with peroxides;
- Weak oil resistance (exclusion aliphatic oils);
- Low resistance to acids, alkalis, and steam.

Applications of silicone rubbers are:

- Roll coverings;
- Molds;
- Seals for the aeronautics industry;
- Lining compounds;
- Technical products and electrical equipment in high temperatures;
- Hospital supplies and medical devices;
- O-rings;
- Cable coverings and insulators.

2.15. POLYSULFIDE RUBBERS (T)

Polysulfide rubbers (PSR) are made when dihalide reacts with sodium polysulfide. PSRs had only a single manufacturer, Morton International. PSRs could be separated into four different groups: ST, FA, Thiokol A, and LP rubbers. A-type PSRs had ethylene dichloride as a dihalide, FA rubbers are formed from the blend of dichloroethylene form and ethylene dichloride. LP kinds are liquid polymers. They are made by breaching down a high molecular weight polymer in a precise manner ST-rubbers are formed from trichloropropane and dichloroethylene form. The sulfur content of A-type is high (84%), The sulfur content of ST types is 37% and FA types are 49% (Figure 2.19) (Sung et al., 2005).

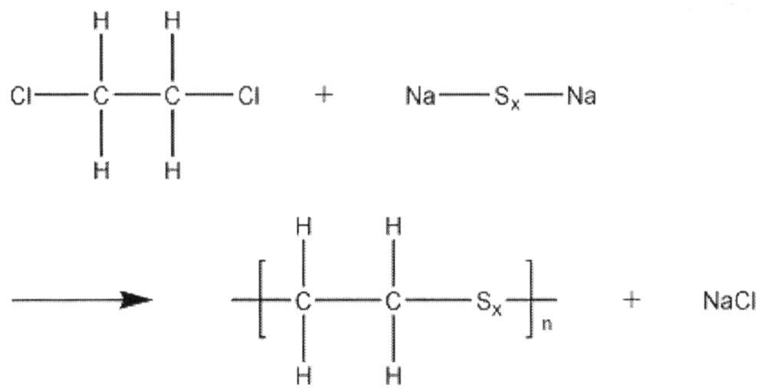

Figure 2.19. The polymerization of polysulfide. Reactants are sodium sulfide and ethyl chloride.

Source: https://docplayer.net/36571488-Elastomeric-materials.html.

The FA and A types are normally vulcanized through the addition of zinc oxide. The LP and ST types are vulcanized with an oxidizing agent, for example, with metal peroxides or metal oxides.

Properties of polysulfide rubbers are:

- Difficult to machine;
- They corrode copper;
- Bad smell;
- Excellent oil and solvent resistance;
- Very good low-temperature properties;
- Good weather and ozone resistance;
- Slight operating temperature range.

Applications of polysulfide rubbers are:

- Varnish and paint rolls;
- Seals;
- Oil, paint, and fuel hoses;
- Roller coverings.

2.16. ETHYLENE-VINYL ACETATE COPOLYMER (EVA)

EVA elastomer is a copolymer of vinyl acetate and ethylene. The properties of the rubber rely on the vinyl acetate content. EVA polymer had rubbery properties when the vinyl acetate content is 40 to 60 wt.% (Figure 2.20) (Faker et al., 2008).

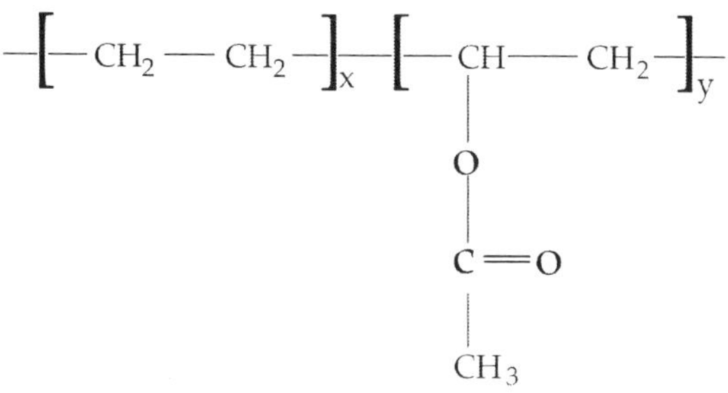

Figure 2.20. Ethylene-vinyl acetate rubber.

Source: https://en.wikipedia.org/wiki/Ethylene-vinyl_acetate.

The technique of preparing ethyl-vinyl acetate relies on the chosen vinyl acetate content. Mass polymerization provides 45 wt.% content at most, emulsion polymerization provides over 50 wt.% content, and solution polymerization 30 to 90 wt.% content. EVA could be vulcanized using ionizing radiation or peroxides. Sulfur can't be used due to the saturated main chain (Maeda et al., 1987).

EVA is often merged with SBR and NR to increase ozone resistance.

Properties of EVA are:

- Fire resistance;
- Good heat resistance;
- Low price;
- Good tack to other materials;
- Excellent ozone, O_2, and light resistance;
- Low abrasion resistance;
- Poor tear resistance;
- Extremely good water and oil resistance;
- No resistance to organic solvents;

- Low elasticity because of the thermoplastic character;
- With reinforcements, high tensile strength could be attained.

Applications of EVA are:

- Floor materials;
- Wire and cable coverings;
- Hoses;
- Seals;
- Some medical extrusions.

2.17. POLYPROPYLENE OXIDE RUBBERS, POLYPROPYLENE OXIDE-ALLYL GLYCIDYL ETHER COPOLYMER (GPO)

Polypropylene oxide rubbers are copolymers of allyl glycidyl ether and propylene oxide. Usually, allyl glycidyl ether content is around 5%. The polymerization technique is solution polymerization in hydrocarbon. Vulcanization could be done with sulfur (Figure 2.21) (Durand, 1986).

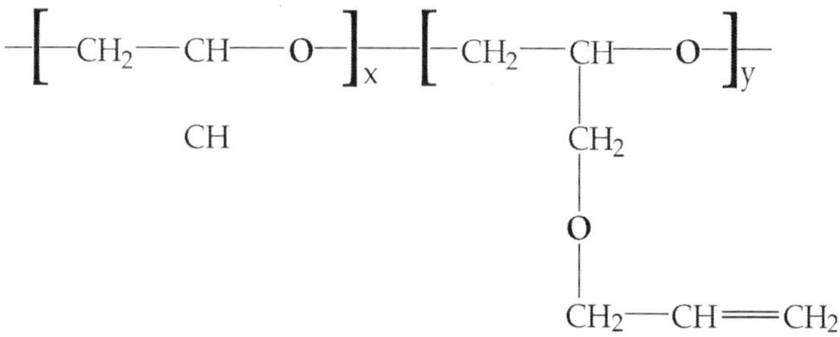

Figure 2.21. Polypropylene oxide rubber.

Source: https://docplayer.net/36571488-Elastomeric-materials.html.

Properties of polypropylene oxide rubbers are:

- Weak oil resistance;
- Low internal damping;
- High price;
- Good heat and cold resistance;
- Good elasticity;

- Good properties at low temperatures;
- Excellent ozone, O_2, and UV light resistance;
- Broad temperature range.

Applications of polypropylene oxide rubbers are:

- Body mounts;
- Suspension bushing;
- Vibration absorbers;
- Seals;
- engine mounts.

2.18. CHLORINATED POLYETHYLENE (CM, CPE), CHLOROSULFONATED POLYETHYLENE (CSM, CSPE)

Polyethylene (PE) is usually a semi-crystalline thermoplastic. But, chlorine could be added to the polymer chain to avoid crystallization. The quantity of chlorine in chlorinated PE decides the properties of the polymer. In utilizing small contents (25%), the material is still crystalline. The amalgamation of higher chlorine content (> 40%) will make the material too brittle. The finest rubbery properties are achieved when chlorine content is around 35%. CSPE is the same as chlorinated polyethylene (CPE); however, it is easier to cure due to the chlorosulfone group. That is why CSPE is utilized more than CPE. The usual chlorosulfone content in the elastomer is fewer than 1.5% (Figures 2.22 and 2.23) (Utracki and Handbook, 2003).

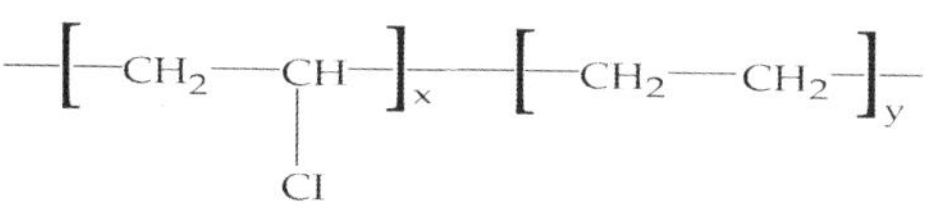

Figure 2.22. Chlorinated polyethylene.

Source: https://polymerdatabase.com/Elastomers/CM.html.

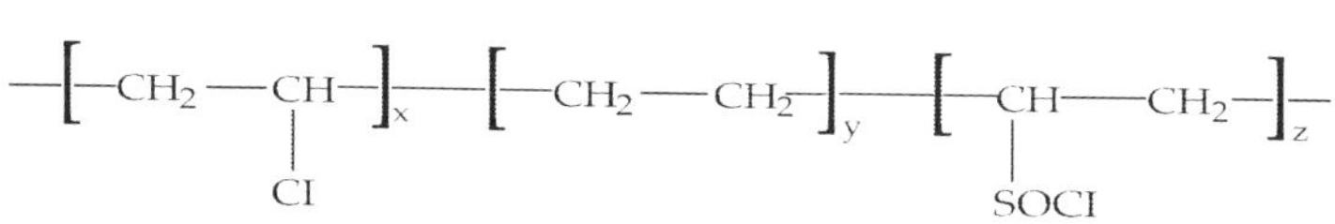

Figure 2.23. *Chlorosulfonated polyethylene.*

Source: http://polymerdatabase.com/Elastomers/Hypalon.html.

CPE could be cured using radiation or peroxides. CSPE could be vulcanized with metal oxides, amines, and peroxides. Enhancing chlorine content enhances fuel, oil, and solvent resistance; however, it reduces low-temperature flexibility (Akovali, 2012).

Properties of CM are:

- Very good dynamic fatigue;
- Good flame resistance;
- Excellent aging resistance;
- Very good UV light resistance;
- Low compression set (up to 150°C);
- Good oil resistance;
- Good flame resistance;
- Very good ozone, O_2, and light resistance;
- Very good color stability;
- Good tensile and breaking strength;
- Very good chemical resistance.

Properties of CSM are:

- High swelling in some types of oils;
- Relatively difficult to process;
- Oxidation and ozone resistance;
- Good cold, heat, and flame resistance;
- Chemical resistance good;
- High compression set in high temperatures.

Applications of chlorinated rubbers are:

- Hoses;
- Automotive tubes;
- Boots;
- Cable and wire coverings;
- Pond liners;
- Floor materials;
- Electrical insulator;
- Dust covers;

- Coated fabrics;
- Molded goods.

2.19. THERMOPLASTIC ELASTOMERS (TPE)

TPEs are a polymer group whose key properties are easy processability and elasticity. The utilization of TPEs had increased noticeably in the latest decades.

TPEs are an extensive group of materials. These materials have numerous advantages of which the most important are (Ganguly and Das, 2016):

- Good chemical resistance;
- Excellent abrasion resistance;
- Easy processability (compared to rubber);
- Damping properties;
- Recyclability;
- Good properties at low temperatures.

Preventive features of TPEs compared to the rubbers are the comparatively low highest operating temperature (< 130–160°C), the high price of TPE, and the small selection of soft grades.

TPEs are utilized in areas where elasticity over an extensive temperature range is needed. The main applications are in the sports accessories and automotive industry.

Thermoplastics elastomers could be divided into the following groups (Juárez et al., 2013):

- Thermoplastic urethane elastomers;
- Thermoplastic amide copolymer, TPE-A;
- Elastomeric alloys;
- Styrene-diene block copolymer;
- Thermoplastic ester-ether copolymers, TPE-E.

2.20. STYRENIC THERMOPLASTIC ELASTOMERS (TPE-S)

Styrenic TPEs comprise styrene/ethylene-butylene copolymer (SEBS), styrene/isoprene copolymer (SIS), styrene/ethylene-propylene copolymer

(SEPS) and styrene/butadiene copolymer (SBS) TPEs found on styrene are block copolymers in which a polydiene unit separates polystyrene blocks. The polydiene might be, for instance, ethylene-butylene (SEBS) or ethylene-propylene (SEPS), isoprene (SIS), butadiene (SBS). The styrene content differs with diverse materials, however, usually, it is 20 to 40% (Figure 2.24) (Spontak and Patel, 2000).

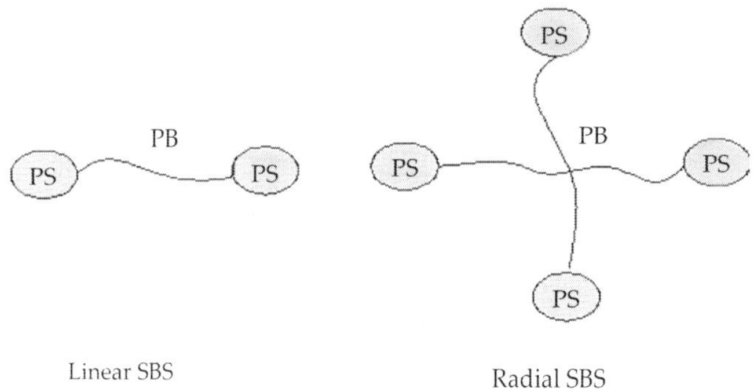

Figure 2.24. The radial and linear structure of styrene thermoplastic.

Source: https://www.mdpi.com/2076-3417/9/4/742.

Advantages of styrenic TPEs are:

- Large variety in hardness;
- Good electrical properties;
- High tensile strength and modulus;
- Colorless, good transparency;
- Good abrasion resistance;
- High friction coefficient (corresponds to that for nr);
- Good miscibility.

Disadvantages of styrenic TPEs are:

- Poor oil and solvent resistance;
- Weak ozone, O_2, and light resistance of SBS (exception SEBS);
- Poor high-temperature resistance (maximum operation temperature, SEBS 135°C, SBS 65°C).

Applications of styrenic TPEs are:

- Adhesives;
- Shoe soles;

- Cables and wires;
- Rubber Products in the car industry;
- With thermoplastics in multi-component injection coextrusion and molding.

2.21. ELASTOMERIC ALLOYS

Elastomeric alloys are mixtures of thermoplastics and elastomers that could be processed utilizing thermoplastic processing techniques. Elastomeric alloys are (Harmer, 2000; Gryshchuk, 2005):

- Thermoplastic Olefin Elastomers (TPO);
- Thermoplastic vulcanizates (TPV);
- Melt Processible Rubbers (MPR).

2.21.1. Thermoplastic Olefin Elastomers (TPO, TOE)

TPO are most frequently blends of Polypropylene and EPDM or Polypropylene and EPM. Butyl rubber and NR had also been utilized. A mixture could be made in a mechanical mixing unit, for example. in polymerization reactors or a twin-screw extruder.

The properties of TPO differ according to mixture ratio, conditions, and components, of alloying. Properties typical of TPO are (Markarian, 2008):

- Low density;
- Low price;
- Good processibility;
- Good chemical resistance;
- Excellent weathering resistance.

Applications of TPO are:

- Wire and cable coatings;
- Hoses;
- Buffers and outside profiles in the car industry.

2.21.2. Thermoplastic Vulcanizates (TPE-V, TPV, DVR)

Thermoplastic vulcanizates (TPVs) are mixtures of elastomers and thermoplastics that had been vigorously vulcanized during their mixing (see picture). Those types of materials are, for instance, dynamically vulcanized mixtures of PP and NBR and PP and EPDM. The properties of the material rely greatly on the content and structure of the elastomer (Figures 2.25 and 2.26) (Spontak and Patel, 2000; Chen et al., 2012).

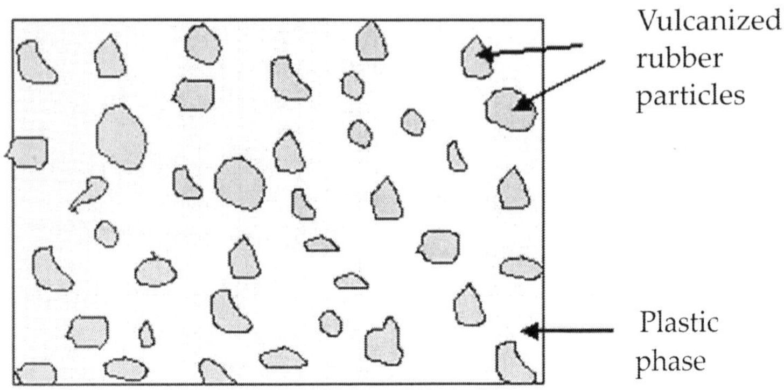

Figure 2.25. The structure of TPE-V, displaying finely dispersed vulcanized rubber particles in the thermoplastics matrix.

Source: https://link.springer.com/referenceworkentry/10.1007%2F978-3-642-29648-2_310.

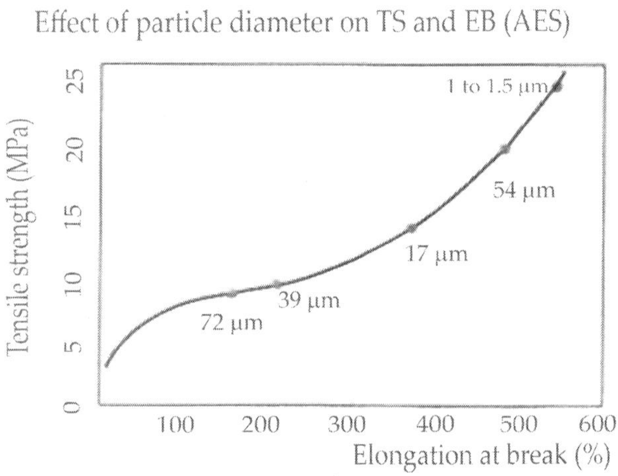

Figure 2.26. The impact of rubber particle size in TPE-V (AES).

Source: http://kompozite.com/2013-09-26/acrylic-rubbers.html.

The properties of thermoplastic vulcanizates:

- Fatigue durability;
- Good properties at low temperatures;
- Good liquid and oil resistance;
- Good mechanical properties;
- Small permanent deformation.

Applications of thermoplastic vulcanizates are:

- Tubes:
- Electrical insulators;
- Car components.

2.21.3. Melt-Processible Rubbers (MPR)

MPR (Melt-processible rubbers) are very rubbery materials that feel and look like traditional rubbers. But, they could be processed like thermoplastics. MPR (Meltprocessible rubbers) had one phase structure, thus they vary from other TPEs that had a two-phase structure (Hamilton and Sahatjian, 1998).

Properties of MPR (melt-processable rubbers) are:

- Flexibility and softness;
- Excellent elasticity;
- Stress-tensile behavior resembles that of vulcanized rubbers.

2.22. THERMOPLASTIC URETHANE ELASTOMERS (TPU, TPE-U)

PUs are termed after the urethane group, which is made when the isocyanate group reacts with the hydroxyl group of the alcohol. Relying on the amount and type of feeding stocks and additives, PUs could be rubbers (PUR), thermoplastics, or TPEs (Figure 2.14) (Habieb et al., 2019). Thermoplastic PU elastomers made from long (MW about 600–3000 g/mol) soft segments of polyethers (TPE-EU) or linear polyester (TPE-AU) and short, hard urethane segments that are made of small alcohol molecule chain extender and di-isocyanate, for example, butanediol (Figure 2.27) (Allen et al., 2000; Norbert, 2012).

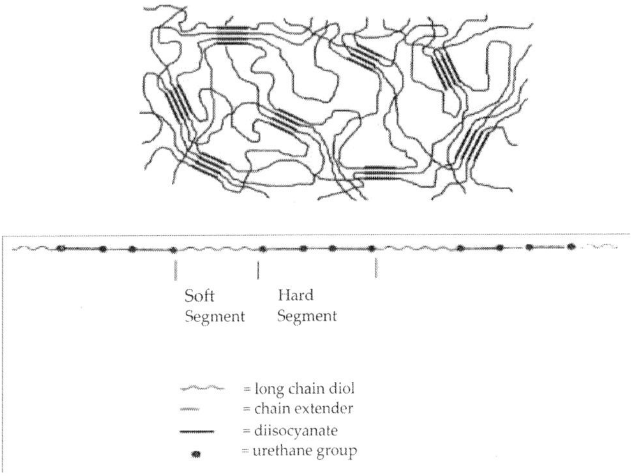

Figure 2.27. The structure of thermoplastic urethane elastomers: Ether diol chains or long ester and hard urethane segments.

Source: https://www.researchgate.net/figure/Chemical-structure-of-PU-elastomer_fig2_324680659.

The properties of thermoplastic urethane (TPU) elastomers differ intensely according to feedstocks and the ratio of soft and hard segments in the material. The soft segment constituent influences particularly the low temperature properties of TPE-U, however also numerous other characteristics. Relying on whether the soft segment is made of polyether or polyester, the properties could be compared according to Table 2.12 (Gur et al., 2014).

Advantages of thermoplastic urethane elastomers:

- Good strength and stiffness properties;
- Good tear strength;
- Low friction coefficient (relying on hardness);
- Good abrasion resistance;
- Good ozone, O_2, and weather resistance.

Disadvantages of thermoplastic urethane elastomers:

- Comparatively poor UV light resistance;
- Poor hydrolysis resistance;
- Poor resistance to aromatic and chlorinated solvents.

Table 2.12. The Properties of TPEU and TPAU

Property	TPEU	TPAU
Tensile strength	0	++
Water absorption	+	0
Tear resistance	0	++
Weather resistance	0	+
Hydrolysis resistance	+	–/0
Low swelling in oil, fat, and petrol	0	+
Abrasion resistance	0	++
Oxidation resistance	–/0	+
Microbes resistance	++	–/0
Radiant energy resistance	0	+
Impact resistance at low temperatures	++/0	0

Note: –: poor, 0: fair, +: good, ++: excellent.

Applications or thermoplastic urethane rubber are (Kim et al., 2008):

- Hoses;
- Cable and wire coatings;
- Components of the car industry;
- Conveyor belts;
- Footwear.

2.23. THERMOPLASTICS POLYESTER-ETHER ELASTOMER (TPE-E)

Polyetherglycols, like polypropylene or polybutylene ether glycols, PE, are soft segments in TPE-Es. Hard segments are: butanediol or dimethyl terephthalate.

Advantages of TPE-E are (Chen et al., 2003; Wilke et al., 2017):

- Good strength properties;
- Good ozone and oxygen resistance;
- Good oil resistance.

Disadvantages of TPE-E are:

- Poor hydrolysis resistance;
- High price;
- Little variety in hardness;
- Poor UV-light resistance; O_2
- Low elongation at break (needs own design principles of products).

Applications of TPE-E are:

- Hoses, tubes;
- Gaskets;
- Cable and wire coatings.

2.24. THERMOPLASTIC POLYAMIDE ELASTOMERS (TPE-A)

Soft segments of polyethers or polyesters and a firm block of polyamide make thermoplastic polyamide elastomers (TPE-A). The polyamide could be, for instance, polycarbonate-ester amide (PCEA), polyetheresteramide (PEEA), polyesteramide (PEA), or polyether-block-amide (PE-b-A). The properties of TPE-A rely strongly on the type of polyol block, the type of polyamide block, and the amount and length of blocks (Figure 2.29) (Langfeld et al., 2015; Wilke et al., 2017).

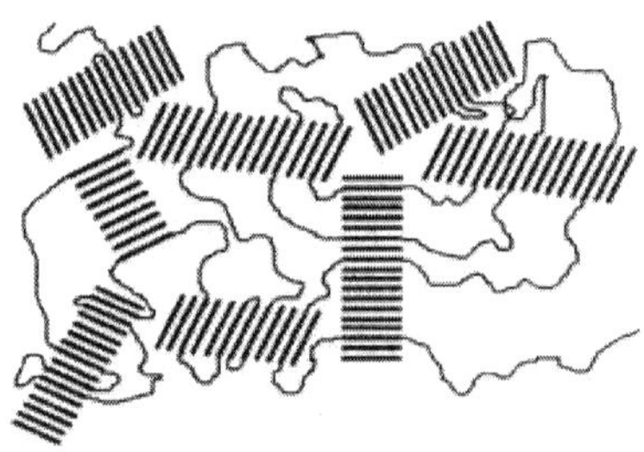

Figure 2.28. The structure of thermoplastic polyamide elastomers.

Source: https://pubs.rsc.org/en/content/articlelanding/2018/py/c8py00068a.

Properties of thermoplastic polyamide elastomers:

- Good abrasion resistance;
- Good chemical resistance;
- Good heat resistance (up to 170°C).

Applications of thermoplastic polyamide elastomers:

- Hoses;
- Components in car motors and under the hood;
- Footballs, skiing boots;
- Films penetrating water vapor;
- Wire and cable coatings.

2.24.1. Comparison of Different TPEs

Certain values for the comparison of different TPEs are given in Table 2.13 (Fielding, 1943; Ma et al., 2003):

Table 2.13. Comparison of Different TPEs

	TPE-E	TPE-V	TPE-U	TPE-A	TPE-S
Density [g/cm^3]	1.1–1.2	0.89–1	1.1–1.3		0.9–1.1
Hardness Shore A/D	40-72D	60A-75D	60A-55D	75-63A	30A-75D
Lowest util T. [°C]	–65	–60	–50	–40	–70
Highest util. T. [°C]	150	135	140	170	70, 135
Compression set at 100°C	F/G	P	F/G	F/G	P(SBS) F/G(SEBS)
Hydrocarbon resistance	G/E	G/E	F/E	G/E	F/E
Hydrolysis resistance	P/G	G/E	F/G	F/G	G/E
Price order [€/kg]	6–8	3–6	4–7	7–10	2–5

2.24.2. New Development Trends Happening in the Field of TPEs

New developments in TPEs comprise (Carothers et al., 1931; Anthony et al., 1942):

- Processing;

- Adhesion and joining;
- Electrical properties, conductivities;
- Material innovations;
- Blends including nanofillers;
- Recycling;
- Product design to maximize the benefits of TPEs;
- Food and health applications, bioapplications;
- New polymerization techniques, metallocene techniques;
- Foamed materials, e.g., supercritical gases;
- Smart products, functionality;
- Paintability;
- Coinjection, coextrusion, over-molding;
- Milling, thermoforming, injection, extrusion, and blow molding (all processing alternatives);
- Product innovations/development, hybrid products;
- Design.

REFERENCES

1. Akovali, G., (2012). Plastic materials: Chlorinated polyethylene (CPE), chlorinated polyvinylchloride (CPVC), chlorosulfonated polyethylene (CSPE) and polychloroprene rubber (CR). In: *Toxicity of Building Materials* (Vol. 1, pp. 54–75). Woodhead Publishing.
2. Allen, N. S., Edge, M., Wilkinson, A., Liauw, C. M., Mourelatou, D., Barrio, J., & Martínez-Zaporta, M. A., (2000). Degradation and stabilization of styrene-ethylene-butadiene-styrene (SEBS) block copolymer. *Polymer Degradation and Stability, 71*(1), 113–122.
3. Amin, A. F., Wiraguna, S. I., Bhuiyan, A. R., & Okui, Y., (2006). Hyper elasticity model for finite element analysis of natural and high damping rubbers in compression and shear. *Journal of Engineering Mechanics, 132*(1), 54–64.
4. Amin, S., & Amin, M., (2011). Thermoplastic elastomeric (TPE) materials and their use in outdoor electrical insulation. *Rev. Adv. Mater. Sci., 29*(2011), 30–115.
5. Aminabhavi, T. M., & Khinnavar, R. S., (1993). Diffusion and sorption of organic liquids through polymer membranes: 10. Polyurethane, nitrile-butadiene rubber and epichlorohydrin versus aliphatic alcohols (C1-C5). *Polymer, 34*(5), 1006–1018.
6. Andersen, S. O., (1963). Characterization of a new type of cross-linkage in resilin, a rubber-like protein. *Biochimica Et Biophysica Acta, 69*, 249–262.
7. Andersson, C., (1999). *Lifespan of Rubber Materials and Thermoplastic Elastomers in Air Water and Oil* (Vol. 1, pp. 1–22). IFP-The Swedish Institute for Fiber and Polymer Research, Mölndal, Sweden.
8. Andrews, R. D., Tobolsky, A. V., & Hanson, E. E., (1946). The theory of permanent set at elevated temperatures in natural and synthetic rubber vulcanizates. *Journal of Applied Physics, 17*(5), 352–361.
9. Anthony, R. L., Caston, R. H., & Guth, E., (1942). Equations of state for natural and synthetic rubber-like materials. I. Unaccelerated natural soft rubber. *The Journal of Physical Chemistry, 46*(8), 826–840.
10. Bahia, H. U., & Davies, R., (1994). Effect of crumb rubber modifiers (CRM) on performance related properties of asphalt binders. *Asphalt Paving Technology, 63*, 414–414.

11. Baker, C. S. L., Gelling, I. R., & Newell, R., (1985). Epoxidized natural rubber. *Rubber Chemistry and Technology, 58*(1), 67–85.

12. Barlow, C., (1978). *The Natural Rubber Industry. Its Development, Technology, and Economy in Malaysia* (Vol. 1, pp. 1–33). Oxford Univ. Press.

13. Bieliński, D. M., Ślusarski, L., Włochowicz, A., & Ślusarczyk, C., (1998). Structure and mechanical properties of nitrile rubbers modified with iodine. *Journal of Applied Polymer Science, 67*(3), 501–512.

14. Bode, H. B., Kerkhoff, K., & Jendrossek, D., (2001). Bacterial degradation of natural and synthetic rubber. *Biomacromolecules, 2*(1), 295–303.

15. Bokobza, L., & Rapoport, O., (2002). Reinforcement of natural rubber. *Journal of Applied Polymer Science, 85*(11), 2301–2316.

16. Boyce, M. C., & Arruda, E. M., (2001). Swelling and mechanical stretching of elastomeric materials. *Mathematics and Mechanics of Solids, 6*(6), 641–659.

17. Capps, R. N., (1989). *Elastomeric Materials for Acoustical Applications* (Vol. 1, pp. 1–33). Naval Research Lab Orlando Fl Underwater Sound Reference Detachment.

18. Carothers, W. H., Williams, I., Collins, A. M., & Kirby, J. E., (1931). Acetylene polymers and their derivatives. II. A new synthetic rubber: Chloroprene and its polymers. *Journal of the American Chemical Society, 53*(11), 4203–4225.

19. Chang, C., & Komoroski, R. A., (1989). NMR imaging of elastomeric materials. *Macromolecules, 22*(2), 600–607.

20. Chen, Y., Kushner, A. M., Williams, G. A., & Guan, Z., (2012). Multiphase design of autonomic self-healing thermoplastic elastomers. *Nature Chemistry, 4*(6), 467.

21. Chen, Z., Christensen, L., & Dahn, J. R., (2003). Large-volume-change electrodes for Li-ion batteries of amorphous alloy particles held by elastomeric tethers. *Electrochemistry Communications, 5*(11), 919–923.

22. Chen, Z., Pei, J., Wang, T., & Amirkhanian, S., (2019). High temperature rheological characteristics of activated crumb rubber modified asphalts. *Construction and Building Materials, 194*, 122–131.

23. Cheng, D. M., Gardner, I. J., Wang, H. C., Frederick, C. B., Dekmezian, A. H., & Hous, P., (1990). Spectroscopic studies of the structures of

butyl and bromobutyl rubbers. *Rubber Chemistry and Technology, 63*(2), 265–275.

24. Chester, S. A., & Anand, L., (2010). A coupled theory of fluid permeation and large deformations for elastomeric materials. *Journal of the Mechanics and Physics of Solids, 58*(11), 1879–1906.
25. Choi, S. S., (2002). Properties of silica-filled styrene-butadiene rubber compounds containing acrylonitrile-butadiene rubber: The influence of the acrylonitrile-butadiene rubber type. *Journal of Applied Polymer Science, 85*(2), 385–393.
26. Ciesielski, A., (1999). *An Introduction to Rubber Technology* (Vol. 1, pp. 1–23). Smithers Rapra Publishing.
27. Danesi, S., & Porter, R. S., (1978). Blends of isotactic polypropylene and ethylene-propylene rubbers: Rheology, morphology and mechanics. *Polymer, 19*(4), 448–457.
28. Delzell, E., Sathiakumar, N., Hovinga, M., Macaluso, M., Julian, J., Larson, R., & Muir, D. C., (1996). A follow-up study of synthetic rubber workers. *Toxicology, 113*(1–3), 182–189.
29. Dondi, G., Tataranni, P., Pettinari, M., Sangiorgi, C., Simone, A., & Vignali, V., (2014). Crumb rubber in cold recycled bituminous mixes: Comparison between traditional crumb rubber and cryogenic crumb rubber. *Construction and Building Materials, 68*, 370–375.
30. Dos, S. D., & Batalha, G. F., (2010). Mechanical behavior characterizing and simulation of polyacrylate rubber. *Journal of Achievements in Materials and Manufacturing Engineering, 38*(1), 33–40.
31. Durand, D., (1986). Network formation-From basic theories towards more realistic models. *Polymer Yearbook, 3*, 229.
32. Eames, W. B., Wallace, S. W., Suway, N. B., & Rogers, L. B., (1979). Accuracy and dimensional stability of elastomeric impression materials. *Journal of Prosthetic Dentistry, 42*(2), 159–162.
33. Faker, M., Aghjeh, M. R., Ghaffari, M., & Seyyedi, S. A., (2008). Rheology, morphology and mechanical properties of polyethylene/ethylene-vinyl acetate copolymer (PE/EVA) blends. *European Polymer Journal, 44*(6), 1834–1842.
34. Fielding, J. H., (1943). Flex life and crystallization of synthetic rubber. *Industrial and Engineering Chemistry, 35*(12), 1259–1261.

35. Findik, F., Yilmaz, R., & Köksal, T., (2004). Investigation of mechanical and physical properties of several industrial rubbers. *Materials and Design, 25*(4), 269–276.
36. Franta, I., (1989). *Elastomers and Rubber Compounding Materials, Manufacture, Properties and Application,* 1, 1–34.
37. Fukuda, T., Fujii, S., Nakamura, Y., & Sasaki, M., (2013). Mechanical properties of silica particle-filled styrene-butadiene rubber composites containing polysulfide-type silane coupling agents: Influence of loading method of silane. *Journal of Applied Polymer Science, 130*(1), 322–329.
38. Gama, D. A., Júnior, J. M. R., De Melo, T. J. A., & Rodrigues, J. K. G., (2016). Rheological studies of asphalt modified with elastomeric polymer. *Construction and Building Materials, 106*, 290–295.
39. Ganguly, S., & Das, N. C., (2016). Chlorosulphonated polyethylene and its composites for electronic applications. In: *Flexible and Stretchable Electronic Composites* (Vol. 1, pp. 229–259). Springer, Cham.
40. Gryshchuk, O., (2005). Commercial condensation and addition thermoplastic elastomers: Composition, properties, and applications. *Handbook of Condensation Thermoplastic Elastomers, 1*, 489–519.
41. Gur, S., Mishra, S. K., & Chakraborty, S., (2014). Performance assessment of buildings isolated by shape-memory-alloy rubber bearing: Comparison with elastomeric bearing under near-fault earthquakes. *Structural Control and Health Monitoring, 21*(4), 449–465.
42. Gürakın, H. K., Turan, A. C., & Deligöz, H., (2020). Synthesis of a novel polyester-ether copolymer and its derivatives as antistatic additives for thermoplastic films. *Polymer Testing, 81*, 106214.
43. Habieb, A. B., Valente, M., & Milani, G., (2019). Hybrid seismic base isolation of a historical masonry church using unbonded fiber-reinforced elastomeric isolators and shape memory alloy wires. *Engineering Structures, 196*, 109281.
44. Hamilton, B., & Sahatjian, R. A., (1998). *U.S. Patent No. 5,797,877* (pp. 1–22). Washington, DC: U.S. Patent and Trademark Office.
45. Hanhi, K., Poikelispaa, M., & Tirila, H. M., (2007). *Elastomeric Materials* (Vol. 1, pp. 1–25). Tampere University of Technology, Tampere.

46. Harmer, A. L., (2000). *Thermoplastic Polyurethane for Optical Cable Jacketing*, 170. EuroCable: EC 2000.
47. Haupt, P., & Sedlan, K., (2001). Viscoplasticity of elastomeric materials: Experimental facts and constitutive modeling. *Archive of Applied Mechanics, 71*(2, 3), 89–109.
48. Henderson, A. M., (1993). Ethylene-vinyl acetate (EVA) copolymers: A general review. *IEEE Electrical Insulation Magazine, 9*(1), 30–38.
49. Horgan, C. O., & Saccomandi, G., (2002). A molecular-statistical basis for the gent constitutive model of rubber elasticity. *Journal of Elasticity, 68*(1–3), 167–176.
50. Hu, Z., Huang, C. Y., & Zhu, L. X., (2008). Study on the properties of acrylic rubber-fluorocarbon rubber blends. *China Elastomerics, 18*(1), 58.
51. Huang, J. J., Keskkula, H., & Paul, D. R., (2004). Rubber toughening of an amorphous polyamide by functionalized SEBS copolymers: Morphology and Izod impact behavior. *Polymer, 45*(12), 4203–4215.
52. Jiang, J., Tang, Q., Pan, X., Xi, Z., Zhao, L., & Yuan, W., (2020). Structure and morphology of thermoplastic polyamide elastomer based on long-chain polyamide 1212 and renewable poly (trimethylene glycol). *Industrial and Engineering Chemistry Research, 59*(39), 17502–17512.
53. Juárez, D., Balart, R., Ferrándiz, S., & Peydró, M. A., (2013). *Classification of Thermoplastic Elastomers Based on Their Reprocess Ability and Elasticity, 1,* 1–22.
54. Kader, M. A., & Bhowmick, A. K., (2003). Thermal ageing, degradation and swelling of acrylate rubber, fluoro rubber, and their blends containing polyfunctional acrylates. *Polymer Degradation and Stability, 79*(2), 283–295.
55. Kim, H. J., Son, C., & Ziaie, B., (2008). A multiaxial stretchable interconnect using liquid-alloy-filled elastomeric microchannels. *Applied Physics Letters, 92*(1), 011904.
56. Kishi, H., Kuwata, M., Matsuda, S., Asami, T., & Murakami, A., (2004). Damping properties of thermoplastic-elastomer interleaved carbon fiber-reinforced epoxy composites. *Composites Science and Technology, 64*(16), 2517–2523.
57. Lacy, A. M., Fukui, H., Bellman, T., & Jendresen, M. D., (1981). Time-dependent accuracy of elastomer impression materials. Part II:

Polyether, polysulfides, and polyvinylsiloxane. *Journal of Prosthetic Dentistry, 45*(3), 329–333.

58. Langfeld, K., Wilke, A., Sut, A., Greiser, S., Ulmer, B., Andrievici, V., & Schartel, B., (2015). Halogen-free fire-retardant styrene-ethylene-butylene-styrene-based thermoplastic elastomers using synergistic aluminum diethyl phosphinate-based combinations. *Journal of Fire Sciences, 33*(2), 157–177.
59. Laura, D. M., Keskkula, H., Barlow, J. W., & Paul, D. R., (2003). Effect of rubber particle size and rubber type on the mechanical properties of glass fiber reinforced, rubber-toughened nylon 6. *Polymer, 44*(11), 3347–3361.
60. Le Gac, P. Y., Le Saux, V., Paris, M., & Marco, Y., (2012). Ageing mechanism and mechanical degradation behavior of polychloroprene rubber in a marine environment: Comparison of accelerated ageing and long term exposure. *Polymer Degradation and Stability, 97*(3), 288–296.
61. Li, Q., Huang, X., Liu, H., Shang, S., Song, Z., & Song, J., (2017). Properties enhancement of room temperature vulcanized silicone rubber by rosin modified aminopropyltriethoxysilane as a crosslinking agent. *ACS Sustainable Chemistry and Engineering, 5*(11), 10002–10010.
62. Lili, Q., Chengya, H., Zhao, H., & Chun, D., (2009). Properties of fluoroelastomer/ethylene-propylene-diene rubber blends [J]. *China Synthetic Rubber Industry, 3, 1–33.*
63. Liu, J., Li, X., Xu, L., & Zhang, P., (2016). Investigation of aging behavior and mechanism of nitrile-butadiene rubber (NBR) in the accelerated thermal aging environment. *Polymer Testing, 54*, 59–66.
64. Ma, X., Chen, J., Li, S., Sha, Q., Liang, A., Li, W., & Feng, Z., (2003). Application of absorption heat transformer to recover waste heat from a synthetic rubber plant. *Applied Thermal Engineering, 23*(7), 797–806.
65. Maeda, A., Hashimoto, K., & Inagami, M., (1987). New polyether elastomers for oil-resistant automotive use. *SAE Transactions, 1*, 695–699.
66. Mahmood, N., Ullah, K. A., Ali, Z., Sohail, K. M., Haq, A. U., Werner, S. K., & Heinrich, G., (2012). Preparation and characterization of thermoplastic polyurethane-urea and carboxylated acrylonitrile-butadiene rubber blend nanocomposites. *Journal of Applied Polymer Science, 123*(6), 3635–3643.

67. Malas, A., Pal, P., & Das, C. K., (2014). Effect of expanded graphite and modified graphite flakes on the physical and thermo-mechanical properties of styrene-butadiene rubber/polybutadiene rubber (SBR/BR) blends. *Materials and Design, 55*, 664–673.

68. Manohar, N., Jayaramudu, J., Suchismita, S., Rajkumar, K., Reddy, A. B., Sadiku, E. R., & Maurya, D. J., (2017). A unique application of the second-order derivative of FTIR-ATR spectra for compositional analyses of natural rubber and polychloroprene rubber and their blends. *Polymer Testing, 62*, 447–453.

69. Manosalvas-Paredes, M., Gallego, J., Saiz, L., & Bermejo, J. M., (2016). Rubber modified binders as an alternative to cellulose fiber-SBS polymers in stone matrix asphalt. *Construction and Building Materials, 121*, 727–732.

70. Markarian, J., (2008). Thermoplastic elastomer compounds continue the upward trend. *Plastics, Additives and Compounding, 10*(5), 38–43.

71. Mars, W. V., & Fatemi, A., (2002). A literature survey on fatigue analysis approaches for rubber. *International Journal of Fatigue, 24*(9), 949–961.

72. Mi, H. Y., Jing, X., Napiwocki, B. N., Hagerty, B. S., Chen, G., & Turng, L. S., (2017). Biocompatible, degradable thermoplastic polyurethane based on polycaprolactone-block-polytetrahydrofuran-block-polycaprolactone copolymers for soft tissue engineering. *Journal of Materials Chemistry B, 5*(22), 4137–4151.

73. Mirhosseini, M. M., Haddadi-Asl, V., & Jouibari, I. S., (2019). A simple and versatile method to tailor physicochemical properties of thermoplastic polyurethane elastomers by using novel mixed soft segments. *Materials Research Express, 6*(6), 065314.

74. Mooibroek, H., & Cornish, K., (2000). Alternative sources of natural rubber. *Applied Microbiology and Biotechnology, 53*(4), 355–365.

75. Mousa, A., & Karger-Kocsis, J., (2001). Rheological and thermodynamical behavior of styrene/butadiene rubber-organoclay nanocomposites. *Macromolecular Materials and Engineering, 286*(4), 260–266.

76. Norbert, V., (2012). Characterization of thermoplastic elastomers by means of temperature scanning stress relaxation measurements. *Thermoplastic Elastomers, 1*, 347–370. Rijeka, Croatia: InTech.

77. Noriman, N. Z., Ismail, H., & Rashid, A. A., (2010). Characterization of styrene-butadiene rubber/recycled acrylonitrile-butadiene rubber (SBR/NBRr) blends: The effects of epoxidized natural rubber (ENR-50) as a compatibilizer. *Polymer Testing, 29*(2), 200–208.
78. Oshinski, A. J., Keskkula, H., & Paul, D. R., (1996). The role of matrix molecular weight in rubber toughened nylon 6 blends: 2. Room temperature Izod impact toughness. *Polymer, 37*(22), 4909–4918.
79. Panigrahi, H., Sreenath, P. R., Bhowmick, A. K., & Kumar, K. D., (2019). Unique compatibilized thermoplastic elastomer from polypropylene and epichlorohydrin rubber. *Polymer, 183*, 121866.
80. Phiri, M. J., Cheruthazhekatt, S., Dimeska, A., & Pasch, H., (2015). Molecular heterogeneity of ethylene-propylene rubbers: New insights through advanced crystallization-based and chromatographic techniques. *Journal of Polymer Science Part A: Polymer Chemistry, 53*(7), 863–874.
81. Poh, B. A., & Ng, C. C., (1998). Effect of silane coupling agents on the Mooney scorch time of silica-filled natural rubber compound. *European Polymer Journal, 34*(7), 975–979.
82. Polgar, L. M., Van, D. M., Broekhuis, A. A., & Picchioni, F., (2015). Use of Diels-Alder chemistry for thermoreversible crosslinking of rubbers: The next step toward recycling of rubber products? *Macromolecules, 48*(19), 7096–7105.
83. Polmanteer, K. E., (1988). Silicone rubber, its development and technological progress. *Rubber Chemistry and Technology, 61*(3), 470–502.
84. Qi, H. J., Joyce, K., & Boyce, M. C., (2003). Durometer hardness and the stress-strain behavior of elastomeric materials. *Rubber Chemistry and Technology, 76*(2), 419–435.
85. Rabani, G., Rosair, G. M., & Kraft, A., (2004). Low-temperature route to thermoplastic polyamide elastomers. *Journal of Polymer Science Part A: Polymer Chemistry, 42*(6), 1449–1460.
86. Sadhu, S., & Bhowmick, A. K., (2004). Preparation and properties of nanocomposites based on acrylonitrile-butadiene rubber, styrene-butadiene rubber, and polybutadiene rubber. *Journal of Polymer Science Part B: Polymer Physics, 42*(9), 1573–1585.
87. Sanders, J. E., Nicholson, B. S., Zachariah, S. G., Cassisi, D. V., Karchin, A., & Fergason, J. R., (2004). Testing of elastomeric liners

used in limb prosthetics: Classification of 15 products by mechanical performance. *Journal of Rehabilitation Research and Development, 41*(2), 1–33.

88. Shankar, R., Ghosh, T. K., & Spontak, R. J., (2007). Dielectric elastomers as next-generation polymeric actuators. *Soft Matter, 3*(9), 1116–1129.
89. Shu, X., & Huang, B., (2014). Recycling of waste tire rubber in asphalt and Portland cement concrete: An overview. *Construction and Building Materials, 67*, 217–224.
90. Simpson, R. B., (2002). *Rubber Basics* (Vol. 1, pp. 1–22). Smithers Rapra Publishing.
91. Spontak, R. J., & Patel, N. P., (2000). Thermoplastic elastomers: Fundamentals and applications. *Current Opinion in Colloid and Interface Science, 5*(5, 6), 333–340.
92. Sung, Y. T., Kum, C. K., Lee, H. S., Kim, J. S., Yoon, H. G., & Kim, W. N., (2005). Effects of crystallinity and crosslinking on the thermal and rheological properties of ethylene-vinyl acetate copolymer. *Polymer, 46*(25), 11844–11848.
93. Teh, P. L., Ishak, Z. M., Hashim, A. S., Karger-Kocsis, J., & Ishiaku, U. S., (2004). Effects of epoxidized natural rubber as a compatibilizer in melt compounded natural rubber-organoclay nanocomposites. *European Polymer Journal, 40*(11), 2513–2521.
94. Tobolsky, A. V., Prettyman, I. B., & Dillon, J. H., (1944). Stress relaxation of natural and synthetic rubber stocks. *Rubber Chemistry and Technology, 17*(3), 551–575.
95. Utracki, L. A., & Handbook, P. B., (2003). *International Abbreviations for Polymers and Polymer Processing* (Vol. 1, pp. 1–32). National Research, Council Canada, Industrial Materials, Institute, Boucherville, QC, Canada.
96. Van, B. J. B., & Poirier, Y., (2007). Establishment of new crops for the production of natural rubber. *Trends in Biotechnology, 25*(11), 522–529.
97. Van, D. J. D., Gnatowski, M., Koutsandreas, A., Burczyk, A., & Duncan, S., (2003). Chemical interaction in blends of polyamide and butyl rubbers. *Journal of Applied Polymer Science, 89*(4), 980–991.
98. Vijayabaskar, V., Tikku, V. K., & Bhowmick, A. K., (2006). Electron beam modification and crosslinking: Influence of nitrile and carboxyl

contents and level of unsaturation on structure and properties of nitrile rubber. *Radiation Physics and Chemistry, 75*(7), 779–792.

99. Von, H. J., Heese, A., Koch, H. U., Peters, K. P., & Hornstein, O. P., (1991). Frequency, spectrum and occupational relevance of type IV allergies to rubber chemicals. *Contact Dermatitis, 24*(4), 244–252.
100. Wang, B., Lu, H., & Kim, G. H., (2002). A damage model for the fatigue life of elastomeric materials. *Mechanics of Materials, 34*(8), 475–483.
101. Wilke, A., Langfeld, K., Ulmer, B., Andrievici, V., Hörold, A., Limbach, P., & Schartel, B., (2017). Halogen-free multicomponent flame retardant thermoplastic styrene-ethylene-butylene-styrene elastomers based on ammonium polyphosphate-expandable graphite synergy. *Industrial and Engineering Chemistry Research, 56*(29), 8251–8263.
102. Wu, Y. P., Wang, Y. Q., Zhang, H. F., Wang, Y. Z., Yu, D. S., Zhang, L. Q., & Yang, J., (2005). Rubber-pristine clay nanocomposites prepared by co-coagulating rubber latex and clay aqueous suspension. *Composites Science and Technology, 65*(7, 8), 1195–1202.
103. Xiao, F., Amirkhanian, S. N., Shen, J., & Putman, B., (2009). Influences of crumb rubber size and type on reclaimed asphalt pavement (RAP) mixtures. *Construction and Building Materials, 23*(2), 1028–1034.
104. Yamasaki, S., (2016). Industrial synthetic methods for rubbers. 8. Polyurethane elastomers. *International Polymer Science and Technology, 43*(11), 29–36.
105. Yang, F., Yao, B., Li, C., Shi, X., Sun, G., & Ma, X., (2017). Performance improvement of the ethylene-vinyl acetate copolymer (EVA) pour point depressant by small dosages of the polymethylsilsesquioxane (PMSQ) microsphere: An experimental study. *Fuel, 207*, 204–213.
106. Yang, X., Li, Q., Li, Z., Xu, X., Liu, H., Shang, S., & Song, Z., (2019). Preparation and characterization of room-temperature-vulcanized silicone rubber using acrylpimaric acid-modified aminopropyltriethoxysilane as a crosslinking agent. *ACS Sustainable Chemistry and Engineering, 7*(5), 4964–4974.
107. Yoon, B., Kim, J. Y., Hong, U., Oh, M. K., Kim, M., Han, S. B., & Suhr, J., (2020). Dynamic viscoelasticity of silica-filled styrene-butadiene rubber/polybutadiene rubber (SBR/BR) elastomer composites. *Composites Part B: Engineering, 187*, 107865.

108. Zhang, H., Wang, Y., Wu, Y., Zhang, L., & Yang, J., (2005). Study on flammability of montmorillonite/styrene-butadiene rubber (SBR) nanocomposites. *Journal of Applied Polymer Science, 97*(3), 844–849.

109. Zhang, P., Zhao, F., Yuan, Y., Shi, X., & Zhao, S., (2010). Network evolution based on general-purpose diene rubbers/sulfur/TBBS system during vulcanization (I). *Polymer, 51*(1), 257–263.

110. Zhao, X., Zhang, G., Lu, F., Zhang, L., & Wu, S., (2016). Molecular-level insight of hindered phenol AO-70/nitrile-butadiene rubber damping composites through a combination of a molecular dynamics simulation and experimental method. *RSC Advances, 6*(89), 85994–86005.

111. Zhong, Y., Wu, W., Wu, R., Luo, Q., & Wang, Z., (2014). The flame retarding mechanism of the novolac as char agent with the fire retardant containing phosphorous-nitrogen in thermoplastic poly (ether ester) elastomer system. *Polymer Degradation and Stability, 105*, 166–177.

112. Zhu, L., Cheung, C. S., Zhang, W. G., & Huang, Z., (2015). Compatibility of different biodiesel composition with acrylonitrile butadiene rubber (NBR). *Fuel, 158*, 288–292.

113. Zhu, P., Zhou, C., Dong, X., Sauer, B. B., Lai, Y., & Wang, D., (2020). The segmental responses to orientation and relaxation of thermoplastic poly(ether-ester) elastomer during cyclic deformation: An *in-situ* WAXD/SAXS study. *Polymer, 188*, 122120.

CHAPTER

3 Elastomeric Blends

Contents

3.1. Introduction 116

3.2. Thermodynamics of Blends of the Rubber 117

3.3. Thermodynamics Aspects Incorporating the Filler Effect in the Rubber Blends 120

3.4. Miscible Rubber Blends 123

3.5. Immiscible Rubber Blends 124

3.6. Compatible Rubber Blends 124

3.7. Incompatible Rubber Blends 125

3.8. Factors Affecting the Performance of Blends of Rubber 125

3.9. Distribution of the Additives of Rubber In Rubber Blends 127

3.10. Preparation of the Rubber Blends 129

3.11. Correlation Between Viscosity and Polarity 132

3.12. Morphology Categorization by Microscopy 133

3.13. Preparation Methods Utilized for Microscopy 133

3.14. Om (Optical Microscopy) of the Rubber Blends 136

3.15. Morphology of the Rubber Blends Through Sem 140

References 146

3.1. INTRODUCTION

Expansion of the industry of rubber has been centered on the aim to make improved rubber products at the minimum expense. Rubbers are often blended in order to get the optimal properties needed for optimal performance of the product of rubber for a quantified span in a specific environment (Ghilarducci et al., 2001; Stephen et al., 2006). The blends of rubber also find their extensive utilization in various technological applications (footwear, tire industry, belting, etc.). Usual rubber included are styrene-butadiene rubber (SBR), natural rubber (NR), and butadiene rubber (BR). The blends of rubber are frequently utilized as ternary or binary blends in the industry of tire in quantified phr (parts per hundred rubber) for expansion of different tire components (tread-base/upper cap, sidewall). Currently, the commercial importance of blends of the rubber is rising bearing in mind the point that the performance of wide-ranging rubbers can be improved by an amalgamation of commercially accessible rubbers to achieve industries' requirements and needs for the high-end performance material (Fujimoto and Yoshimiya, 1968; Zhang et al., 2001). Dissimilar rubbers when blended, leads to a range of morphologies, which normally can disturb the end-utilization performance of a product. Therefore, optimizing parameters in order to stabilize the morphology, that is, consistency of the dispersion in batch to batch and inside the batch is of main significance while processing rubber blends. The examinations of the morphological characteristics of the blends of rubber are significant to the large amount as the morphology governs the performance and property of the blends (Botros et al., 2006; Maria et al., 2014).

The terms morphology and structure are utilized interchangeably while recording outcomes for microscopy, but it varies in comprehension, which has been explained by Linda and the co-workers in the book-Polymer Microscopy (Figure 3.1). The word morphology in the rubber compound exhibits the shape, size, and steadiness of the slight rubber phases, additives, and fillers, the relation and size distribution of additives/structural units, and fillers inside the macromolecular structure (Sawyer and Grubb, 1996). The microscopic techniques for the rubber technologist are significant tools in studies associated with NR and synthetic blends and the synthetic blends of rubber; as it aids to examine and analyze morphological changes and complex structures that specific rubbers in the rubber blends show on multiple scales of length (Michler, 2008; Wang et al., 2012).

There is some wide-ranging microscopic method available to perceive and study the microstructure and morphology of the rubber blends. Optical microscopy (OM), scanning electron microscopy (SEM), atomic force microscopy (AFM), and transmission electron microscopy (TEM) are equipped to envisage morphology, filler migration, and phase distribution in the rubber blends. SEM and OM are widely utilized to disclose the microstructures and macrostructures in the macromolecules, respectively. TEM and AFM are utilized for the determination of nanostructures and microstructures in the blends of rubber (Das et al., 2008). Modern microscopes are incorporated with systems that provide local chemical information adding to the structural image. Combining several methods of microscopy with spectroscopy provides the best understanding of the rubber blend morphology. Confocal Raman AFM is the spectroscopic and microscopic analysis technique for the filler migration and local phase morphology analysis. In this present review, the in-depth examination of different sides of rubber blends and the morphological characterization utilizing microscopic techniques is provided (Botros, 2002; Zhang, 2009).

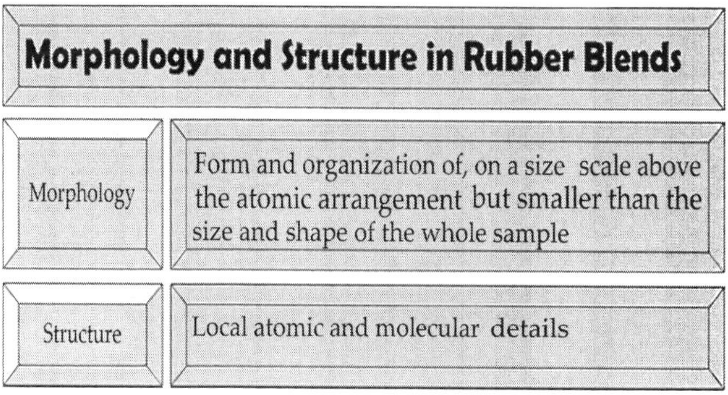

Figure 3.1. Structure and morphology in rubber blends.

Source: https://link.springer.com/chapter/10.1007/978-94-015-8595-8_6.

3.2. THERMODYNAMICS OF BLENDS OF THE RUBBER

Before going into a description for the thermodynamic understanding of rubber/filler-rubber mixture, it is essential to have a brief introduction to the thermodynamics of binary blends of rubber; therefore, a brief on the thermodynamics indulged in the blends of rubber are defined below (Sirqueira and Soares, 2002; Pandey et al., 2005).

Morphology of the blend of rubber relies upon the creation methods utilized and on the kind of additives in the blend of rubber. The morphology of blends of the rubber is disturbed by the mixing level between the two rubbers. The thermodynamic miscibility of the blend is dependent on the composition and temperature. Figure 3.2(a)–(c) displays 3 phase diagrams for the blends of rubber at different temperatures and compositions. Y and X axes display the temperature of rubber blends and composition, respectively (Asakura and Ando, 1998; Sahakaro et al., 2007).

In Figure 3.2(a), the blend turns out to be immiscible beneath the solid line and is called the upper critical solution temperature (UCST). UCST in the rubber blends is observed rarely. If the blend turns out to be immiscible above the given solid line, then the behavior of a rubber blend is lower critical solution temperature (LCST), as displayed in Figure 3.2(b). In some of the blends, LCST and UCST occur instantaneously, as revealed in Figure 3.2(c).

Thermodynamically, blends of rubber can be categorized as displayed in Figure 3.3. Miscible blends are the ones where morphology is similar on the segmental level. Immiscible blends of rubber are those that are heterogeneous and create two distinct components of specific rubber upon blending. The miscible blends of rubber are very rare due to the rubber's high molecular weight. The motive for miscibility can generally be described thermodynamically by the equation below (El-Sabbagh, 2003; Pierson et al., 2010):

$$\Delta G = \Delta H - T\Delta S$$

Where:

ΔG: Gibbs free energy;

ΔH: Enthalpy change;

ΔS: Variation in entropy of mixing.

For the miscible rubber blend, variation in free energy on mixing should be negative, TDS comprises combinatorial mixing entropy, which is dependent on the number of the molecules in a system and is small for the molecular weights related with rubbers and TDS overall becomes the negative value. This consequences in the loss in volume on the mixing of the two rubbers (Saad and El-Sabbagh, 2001; Cappella et al., 2004).

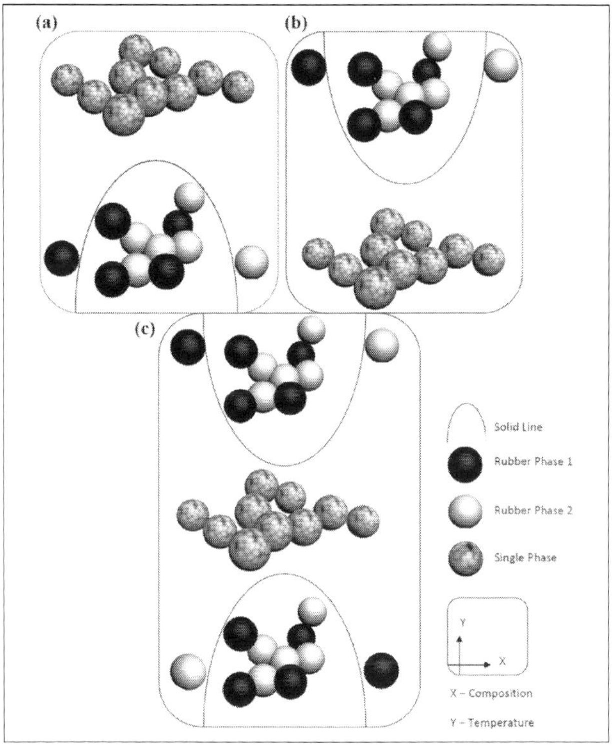

Figure 3.2. Phase image for the blend of rubber, displaying: (a) UCST; (b) LCST; and (c) UCST þ LCST.

Source: https://www.elsevier.com/books/solid-state-nmr-of-polymers/asakura/978-0-444-82924-5.

The miscible blends of rubber will be ascending only if the change in free energy is negative by a feature of the exothermic heat of mixing (ΔH). TΔS is generally small for the low molecular weights linked with rubbery polymers; it provides a negative effect on the mixing of the two rubbers (Joseph et al., 1988).

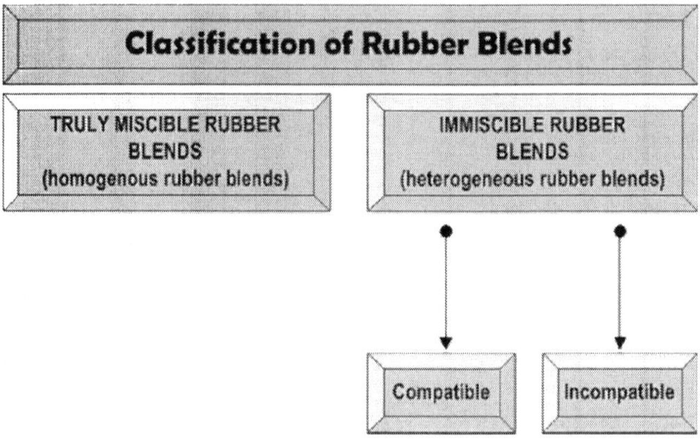

Figure 3.3. Cataloging chart of the rubber blends.

Source: https://www.elsevier.com/books/solid-state-nmr-of-polymers/asakura/978-0-444-82924-5.

Term entropy provides the path for immiscible blends of rubber-rubber. The process of mixing disturbs the morphology and miscibility of the polymer blends (Stephens and Bhowmick, 2001). The blends of rubber are classified as displayed in Figure 3.3. The miscible blends are categorized by thermodynamic factors, whereas compatible blends are categorized by ultimate properties and fine morphology. The miscibility must be devised with thermodynamic, and compatibility must be associated with fine/uniform morphology and good technological properties. Thus miscible blends of rubber must not be compatible from the technological viewpoint. Fascinatingly, most of the industrially significant blends of rubber/rubber are immiscible centered on the thermodynamic considerations; though, these blends are compatible in the technological sense. Frequently, compatibilizers are normally added to the incompatible blends in order to make them compatible blends for industrial applications (Naskar et al., 1994; Naseri and Jalali-Arani, 2015).

3.3. THERMODYNAMICS ASPECTS INCORPORATING THE FILLER EFFECT IN THE RUBBER BLENDS

For a conversation on the thermodynamic aspects comprising fillers, particularly nanofillers, knowledge of the behavior of the thermodynamic way out is essential. Miscibility of the two components is governed by the Eqn. (1) as stated in the section above on thermodynamics for the rubber blends. Enthalpy and entropy during mixing control the dispersion of the fillers in a rubber matrix and lastly, the structure of composites of rubber (Mutz, 2016). In the circumstance of rubbers, because of the existence of tangled long chains, the filler particles aren't able to upsurge the mixing entropy. Therefore, to upsurge negative free mixing energy, intermolecular interactions during the mixing must be tuned. Several articles were published in the past, along with the present outline chain wrapping and chemical functionalization in order to make particles of filler thermodynamically miscible in a polymer solution. But the two methods are accountable for surface modification of the filler particles and disturb filler properties. Therefore, identification of a suitable solvent that forms favorable enthalpic collaborations for the negative free mixing energy is anticipated. The theory of Flory Huggins Solution helps to understand the competence of the solvent in dispersing the particles of filler; a statistical calculation centered on this theory associates entropy and enthalpy of mixing to molecular properties and free energy of mixing.

$$(\Delta G_{mix}/RT)\ [\text{entropic}] = (\phi_A/M)\ ln\ \phi_A[\text{entropic}] + (\phi_B/M)\ ln\ \phi_B[\text{enthalpic}] + \chi_{AB}\ \phi_A\ \phi_B \quad (2) \tag{2}$$

Eqn. (2) provides the equation of Flory Huggins Solution Theory. A system having the value of equation > 0.5 is because of ΔG > 0 and positive ΔH, therefore, creating immiscible systems, where the repulsive interactions amongst components of the systems are far more than the change in entropy inside the system. Therefore, to accomplish the miscibility condition, the value of the equation must be negative or small. The value of a parameter of Hildebrand solubility gives the degree of interaction amongst the components of a system, and the components with comparable values are possible to be miscibly succeeding the norm of like dissolves like. Getting into the details of thermodynamics of the filled blends of rubber, a theory of attraction-depletion is the reason for the existence of aggregates in the composites of a polymer. According to the theory of attraction-depletion, lack of enthalpic interactions inflicts entropic restrictions on a polymer because of the existence of fillers and this triggers excessive phase separation into filler-rich and polymer-rich phases (Fenouillot et al., 2009). Amongst the revolutionary works executed by Mackay et al. in which they validated that R_p/R_g (R_g → radius of gyration for the polymer chains; R_p → radius of a particle of the filler) is a vital parameter in envisaging the dispersion of the fillers. Filler particles with a radius larger as compared to R_g tend to be combined and those having radius smaller than the R_g are well-dispersed. The well-dispersed fillers swell the molecules of the polymer thus increasing their efficient R_g (de Luzuriaga, 2008). Further research reports slowing down of the phase separation by permanent particles of filler, as they accredit to the formation of obstacles for the growth of interface, which triggers saturation of the size of the domain. Mobile particles slow down the process of phase separation but aren't as efficient as immobile particles. The particles of filler having attraction towards one rubber phase and the fillers with anisotropy normally slow down the process of phase separation. Filler particles with an attraction towards both of the rubbers normally will be gathering at interfaces. The filler particle distribution between the bulk phase and the interface can be decided by the Boltzmann factor (Ginzburg, 2010). The localization of filler in the rubber blends is an important aspect of the thermodynamics of the filled blends of rubber. The morphological development occurs by the existence of filler in the rubber blends. The property of the final blend is dependent on the filler localization, dissemination, and creation of the filler networks in a rubber blend system. The localization of filler in the rubber blends is primarily dependent on interfacial energy amongst the filler and rubbers (Bischoff and Scanlon, 2007). The wetting parameters can generally be utilized to envisage the filler localization and interfacial energies is calculated utilizing Eqn. (3) (Mondal and Khastgir, 2017):

$$\omega_{12} = \frac{\gamma_{s-2} - \gamma_{s-1}}{\gamma_{12}} \tag{3}$$

where; γ_{s-1} and γ_{s-2} are the interfacial tension amongst filler and rubbers 1 and 2; g_{12} is the interfacial tension between the two rubbers.

When $1 < \omega_{12} < 1$, the particle of filler will normally be localized at interface amongst rubbers 1 and 2. The upsurge in filler loading upsurges the interfacial area because of a reduction in the size of phases of the dispersed rubber phase. In the system of rubber blend of a_1 and a_2 comprising of the filler d, the equation of free energy of mixing is written as (Vo and Giannelis, 2007):

$$\Delta G = \Delta G_{\alpha 1}\delta + \Delta G_{\alpha 2}\delta - \Delta G_{\alpha 1,\alpha 2} \tag{4}$$

The system is usually stable when the $\Delta G < 0$. The rubber blend is immiscible when the $\Delta G_{\alpha 1,\ \alpha 2}$ has the positive value and it is miscible when $G_{\alpha 1,\ \alpha 2}$ is greater than zero and $G_{\alpha 1}\ \delta$ and $G_{\alpha 2}\ \delta$ is less than 0. If a filler gets adsorbed on the 2 phases of rubber a_1 and a_2, then it behaves like a compatibilizer. ΔG becomes negative because of the rubber-filler collaboration (Figures 3.4 and 3.5) (Ray and Bousmina, 2006).

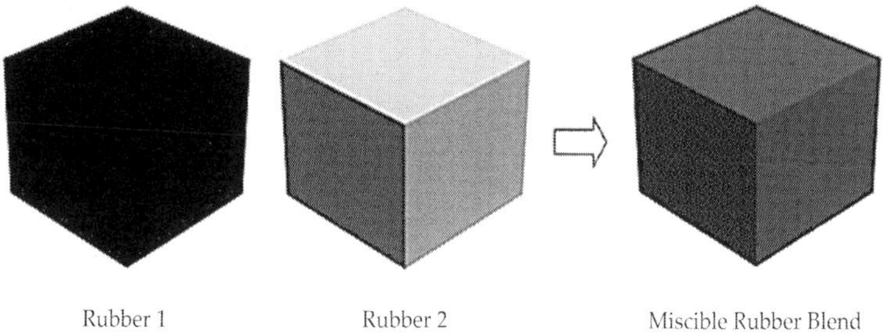

Figure 3.4. Anticipated representation for the miscible rubber blends.

Source: https://www.worldcat.org/title/handbook-of-elastomers/oclc/45247371.

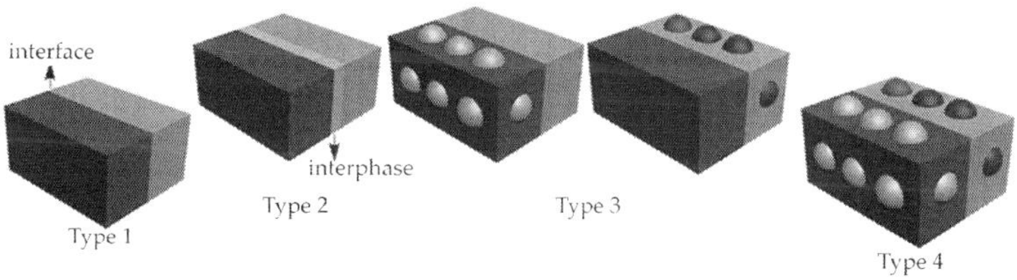

Figure 3.5. Schematic representation for the immiscible blends.

Source: https://journals.sagepub.com/doi/abs/10.1177/1477760619895002.

3.4. MISCIBLE RUBBER BLENDS

The blends of rubber whose mixing free energy gives the negative value are miscible blends of rubber. Miscibility is the thermodynamic term and it infers the homogenous mixture on the level of an atom. Some instances of miscible blends of rubber include SBR of several styrene levels, NR with vinyl BR, Acrylonitrile BR of several acrylonitrile contents, and so on. The existence of miscible blends of rubber-rubber is rare. Miscibility generally upsurges with temperature (Hansen, 1967). Given below is the thermodynamic condition for a miscible rubber blend system.

$\Delta G = \Delta H\ T\Delta S$, for the miscible system ΔG, is less than zero.

$(\Delta G_{mix}/RT)$ [entropic] = $(\Phi_A/M) \ln \Phi_A$ [entropic] + $(\Phi_B/M) \ln \Phi_B$ [enthalpic] + $\varkappa_{AB}\ \Phi_A\ \Phi_B$, for the miscible system must be negative or small.

3.5. IMMISCIBLE RUBBER BLENDS

Blends of rubber whose mixing free energy gives the positive value are generally immiscible blends of rubber. They display poor adhesion and coarse morphology between the two components. These rubber blends are also known as heterogeneous blends. PBR (1,4 Polybutadiene rubber)/NR blend is an instance of the heterogeneous blend. The idea of immiscibility in the polymers has been given in a book named Solid-State NMR of the Polymers, as displayed in Figure 3.5. In type one, the phase domain of just one part of the polymers is disseminated in the matrix of the second polymer having a clear borderline; In type two, polymer chains of the two components are mingled and seem as interphase. If the interphase in type two is higher than the blend becomes the homogeneous blend. Whereas if the interphase is relatively small then type two develops into type one. When one of the polymer parts is semi-crystalline and the other is amorphous, there happens the homogeneous mixing of 1 polymer part with the second part. Type 4 signifies the coexistence of 2 homogenous domains with diverse compositions and is because of the thermal tempted phase separation (Frisch et al., 1970).

3.6. COMPATIBLE RUBBER BLENDS

Incompatible and compatible blends of rubber form the part of immiscible blends of rubber. In most of the blends of rubber, the components aren't molecularly disseminated, that is, the phase isolated state of mixing happens at the molecular level with composition of the isolated phases, pure or alike pure components before blending (as observed in Figure 3.6(b)). They might be denoted as compatible. It might be more suitable to give the term compatible in order to define the systems that do not instantaneously de-blend on the visible scale. The morphology of compatible blends is dependent upon methods of blending and the rheological properties of blend components along with the thermodynamic observations. The compatible structure can generally be the dispersion of 1 phase in the continual matrix of others or both of the phases can be tenacious. Co-continuity infers an interpenetrating polymer network (IPN). Frisch et al. (1970) determined that IPN can normally be intentionally made by synthesis or by governing mixing methods.

3.7. INCOMPATIBLE RUBBER BLENDS

The term incompatibility gives the mean that when two or more than two components of composition don't blend properly to yield a homogeneous or uniform mixture. In the incompatible blends, normally properties are poor than the individual rubbers (Gooch, 2010). This kind of rubber might be made compatible simply by the addition of the compatibilizers like diblock copolymers, interfacial crosslinkers, graft copolymers, and nanofiller (as displayed in Figure 3.6(a)).

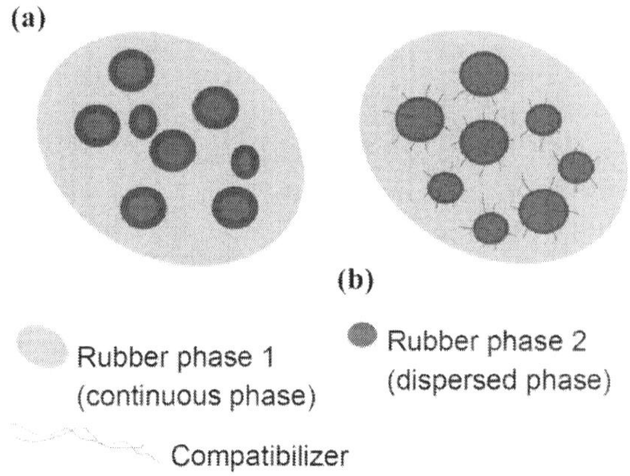

Figure 3.6. Schematic diagram of (a) incompatible blends of rubber-rubber; and (b) compatible blends of rubber-rubber.

Source: https://journals.sagepub.com/doi/abs/10.1177/1477760619895002.

3.8. FACTORS AFFECTING THE PERFORMANCE OF BLENDS OF RUBBER

The blending of dissimilar rubbers is a very complicated process. Miscible rubber blends are uncommon, as mentioned earlier. The major factors which influence the performance of blends of the rubber are polarity, mixing conditions, viscosity ratio, the chemistry of filler, and temperature and are summarized in Table 3.1 with respect to their significance and examples.

Table 3.1. Factors Affecting the Performance of Blends of the Rubber

Factor	Significance	Example
Polarity	The difference in polarity reports interfacial energy. Higher the difference of higher the phase size and interfacial tension.	NBR and NR have dissimilar polarities and have a high difference in their solubility parameters occasioning very high interfacial tension.
Mixing conditions	Mixing conditions have an impact in governing the phase size of the rubber components in blends of rubber. Rubber blends treated at very high shear during the process mixing causes strong interfaces amongst two blending rubbers, also cross-links amongst two rubbers upsurge and blends treated at low shear causes weak interfaces. Deformation of the dispersed rubber domains occurs in high shear parts of the mixing equipment, and if stated processing conditions are kept, then the rubber domains get cracked and develop smaller particles.	Roland and Bohm (1985) discovered that the blend morphology acquired signifies the competition amongst the dispersion of rubber particles and the flow-induced coalescence.
Viscosity	The resemblance in viscosity outcomes in co-continuous morphology (Sae-Oui et al., 2016). Viscosity of only one phase can be upsurged by the addition of the cross-linking agents to one phase.	The natural rubber has high preliminary viscosity as compared to the other rubbers. During the mixing process, molecules of NR break down quickly which will yield a mismatch in the viscosities and it does not provide good phase morphology.
Rubber blend ratio	The optimal quantity of the rubber components in a blend of rubber must be monitored. Size and type are the two sides of phase morphology. Normally, one phase is detached inside another phase, or either it is co-continuous be contingent on the polymer ratio. Size denotes the domain size of the rubber phases in the blend.	
Temperature	The miscible blends of the polymer cause phase separation upon the upsurge in temperature. The temperature at which curing or blending is carried out might be greater as compared to the lower crucial solution temperature of the rubber pair. Phase isolation of the miscible polymers at higher temperatures outcomes principally from an upsurge in the loss of entropy linked with the volume variations accompanying mixing.	Studies by Inoue et al. (1985) stated the upper crucial solution temperature in styrene-butadiene rubber-BR blends.

Abbreviations: SBR: styrene-butadiene rubber; NR: natural rubber; NBR: acrylonitrile-butadiene rubber.

3.9. DISTRIBUTION OF THE ADDITIVES OF RUBBER IN RUBBER BLENDS

Compounding is the material science of adjusting rubber-rubber and rubber blends in order to meet the necessity during service performance (Donnet and Custodero, 2005). Several compounding ingredients for the modification of rubber include stabilizers, fillers (antiozonant and antioxidant), special additives, and curing package (accelerator, sulfur, and activators). Compounding is performed on the two-roll mill or executed on the internal mixer, succeeding in a decrease in viscosity, dispersion, distribution, and incorporation (De and White, 2001). The ingredients of compounding are classified as inactive ingredients and active ingredients; some categorize them as insoluble ingredients and soluble ingredients. Carbon black (CB) is an ingredient loaded in very large quantity, impacts (decline or enhancement) the properties of rubber-rubber and rubber blends compounds, because of its chemical collaborations with the chains of rubber (physical adsorption on CB surface), and creates the 3-D network. Apart from CB, which is existent in very large quantities-cross-links are formed by sulfur. These two ingredients have a major role despite their quantity in the compounding formulation. CB is an inactive and insoluble ingredient. Sulfur is a reactive and soluble ingredient. Comprehensive information on distribution features of sulfur and CB is mentioned below from the available literature available. Miscibility and compatibility in the rubber-rubber blends are two words which are utilized on behalf of one another, but miscibility in the rubber-rubber blends conforms to molecular weight and the intermolecular interactions inside the components of blends of the rubber-rubber; whereas compatibility in technological logic is utilized to define whether the anticipated result happens when the two polymers are blended. Compatibility and miscibility for the filled blends are impacted by the partition of filler amongst the phases and dispersion in the separate phases (Isayev and Palsule, 2011).

The extent of the affinity of the rubber towards fillers is a significant factor in concluding the partition of filler. The chains of rubber are adsorbed physically on the surface of filler, till the adsorption enthalpy is obtained owing to the privileged affinity between filler surface and rubber segments, and therefore, filler partition is controlled by all of the factors which control the adsorption of chains of rubber onto the surfaces of filler. Without leading to the equilibrium state, the outcome of this distribution procedure can direct the balance of filler-polymer interactions. Additionally, the procedure of mixing and the conditions of the process employed can substantially impact the kinetics of the partition of filler. Therefore, the tendency of the partition of filler can be anticipated from competing for adsorption of polymers on the fillers. The phenomenon adsorption of the chains of rubber on the rubber chains in the blends of rubber-rubber has to be well-thought-out for two kinds of rubber-rubber blends. The blends of rubber-rubber having high interfacial tension with intensely differing polarity. The polarity difference causes the privileged adsorption of one of the rubber components on the surface of filler, despite morphology having large domains, which validates fillers partitioning quantitively to the most interrelating rubber within the blend of rubber-rubber. Secondly, the blends of rubber-rubber made compatible due to their resemblance in the solubility parameter and also possess small interfacial tension, comparable affinity to filler; in which the partition of filler becomes more balanced (Limper, 2012).

In blends having the different polarity that is created from the unsaturated chain backbone and saturated chain backbone, a preferred distribution of CB into an unsaturated polymer is observed always. In the rubber-rubber blend of NR and chloro-butyl rubber, CB particles are mostly positioned in the NR phase and to some degree in the neighborhood of phase boundary amongst the chloro-butyl rubber and NR, while there is almost no CB in a chloro-butyl rubber. In one more study with BR and chloro-butyl rubber blends, CB favors the BR phase. The experiments with NR blends demonstrate that the dissemination of CB's resembles qualitatively the chemical-physical affinity between the rubber and the filler. Favored content of CB is always existent in rubber with a higher extent of unsaturation when the blends of chloro-butyl, butyl, or ethylene propylene diene monomer (EPDM) rubber with the diene rubbers are created. In blends developed from non-polar rubber and polar rubber, a preferred distribution of the polar fillers to the polar rubber is witnessed. Irrespective of viscosity ratio and viscosity in blends of the acrylonitrile-butadiene rubber (NBR) EPDM and BR, silica is disseminated quantitatively in NBR. Noticeable transformations can take place in the partition when the rubber constituent's nature becomes analogous; that is, the contentment of double bonds or explicitly interacting groups. The scientific reports validate fillers partition, in the blends of rubber-rubber is dependent on molecular weight, that is, the viscosity of rubber and to some extent are impacted by the filler surface specific area. CB partition in the blends of NR with BR and SBR was repeatedly discovered to be superior in the BR and SBR phase. Additionally, it was validated that CB favorite for the styrene-butadiene rubber phase is because of the attraction of phenyl groups (C_6H_5)

to the sites of adsorption of the CB (Limper, 2012). Fillers dispersion is the most frequently considered issue in the area of rubber. Carbon black distribution is monitored by making well-dispersed batches of every rubber component with stated loading. Figure 3.7 displays the even distribution of CB amongst the two rubber phases.

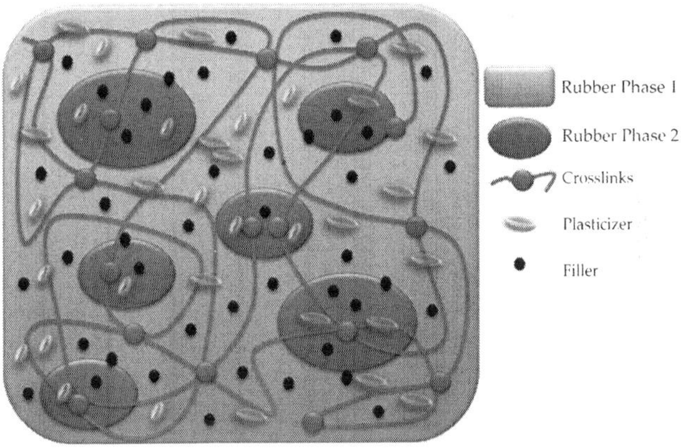

Figure 3.7. Pictorial diagram of the cross-linked usual blend of rubber with additives.

Source: https://journals.sagepub.com/doi/abs/10.1177/1477760619895002.

The soluble ingredients distribution, that is, a benefits package, impacts the performance of vulcanized rubber products. The diffusion of tetramethyl thiuram disulfide, sulfur, diphenyl guanidine, and benzothiazole disulfide takes place from unvulcanized compounds of rubber of the low unsaturated elastomers (like EPDM and butyl rubber) to elastomers with high unsaturation (like SBR and NR). The curatives relocation was detected to occur very quickly. The gradient of diffusion, before vulcanization takes place, might be produced amongst the elastomers of different nature. The curative disproportion amongst the constituent phases and associated under cure and over cure are the practical outcomes for diffusion differences and solubility in the rubber-rubber blends (Fujimaki and Oshima, 1989).

3.10. PREPARATION OF THE RUBBER BLENDS

Preparation methods, with their comprehensive features, benefits, and drawbacks with examples are given in Table 3.2.

Table 3.2. Different Methods of Preparation for Rubber Blends

Preparation Technique	**Characteristics**	**Example**	**Advantages/Disadvantages**
Latex blending	Blending with the latex stage provides finer scale dispersion as compared to solution blending. Latex blending also has the capability for subtle dispersion of the specific latex particles, as the particles in latex are fine and are well-dispersed with the assistance of surfactants (Sui et al., 2008).	Japan SR has the patent which declares that NR blends with SBR or BR, when made by latex blending provides the homogeneous dispersion of carbon black.	Give a good degree of reinforcement and dispersion. Environmentally harmless as they do not utilize any solvents (Jose and Joseph, 2005).
Dry blending	Made by mechanical mixing with extreme shear, having beneficial properties, and are homogenous macroscopically. Extruders, two roll mills, and internal mixers are the equipment of mixing which provides needed shear force in order to improve blend formation.	Mixing of EPDM and NR was performed in the internal mixer at nearly 60 rpm/min at 50 °C (Chang et al., 1999; Rathi et al., 2018).	The process of dry blending is economical and allows to integrate of compounding ingredients in just one step. The costs of dry blending are very high as it needs costly machinery and man Power (Corish, 1967).
Solution blending	Dissolution of the two rubbers in an appropriate solvent occurs and it gives rougher particles because of the low viscosity of the solution.	XNBR and NR solutions can be made by chloroform or toluene (Satyanarayana et al., 2016).	Rapid elimination of solvent seems to be the suggested procedure to decrease heterogeneity, especially because of black redistribution during drying.
Blending with the powdered rubbers	Blending of 2 rubbers in the form of powder is also Permissible (Mangaraj, 2004).	Wang et al. had developed UFPSBR, which normally has been mixed with Natural rubber/SBR in the two-roll mill (Liu et al., 2012).	Powdered rubbers are easier in handling so process loss is less. The rates of Production are higher when utilizing powdered rubber and the 2-stage mixing can generally be removed. Less energy is required for the processing of powdered rubber.

Mechano-chemical blending	Reactive monomers normally when added to both or one of the components, give better blends because of *in situ* creation of block or graft copolymer that behaves as a compatibilizer.	Block copolymers are made when the radicals of the polymer tend to recombine instead of disproportionate. Inter-polymerization occurs in blends of NR, chloroprene rubber, styrene-butadiene rubber, butadiene rubber, and nitrile butadiene rubber when mastication is executed under a nitrogen atmosphere (Yehia et al., 2004).	Accurate control of the rubber blend structure provides great promise for new properties and applications.
Freeze drying	Amorphous materials generally solidify over the temperature range known as the freezing range. The mass formed after the process of freeze-drying can usually be excellently ground to powders appropriate for electrostatic spraying and the other powder procedure for product development (Ifuku et al., 2011).	–	The major benefits of freeze-drying are the ingredients of freeze-dried material stay homogeneously dispersed. It does not cause toughening or shrinkage of the material dried, and the ultimate product attained will be lightweight. The product is kept in a dry state, so there isn't any stability issue. The method is expensive as the dryers are costly. Freeze-dried samples frequently exhibit the non-uniform morphology as the chains of rubber in a solution don't get adequate time for orientation, and the distribution of molecular weight becomes non-uniform (Li and Cho, 2017).

Abbreviations: XNBR: carboxylated nitrile butadiene rubber; NR: natural rubber; UFPSBR: ultra-fine full vulcanized powdered styrene-butadiene rubber; EPDM: ethylene propylene diene monomer.

3.11. CORRELATION BETWEEN VISCOSITY AND POLARITY

In the two-phase blend of polymer with scatter matrix morphology, the ratio of the viscosity of a separate phase to a matrix plays a significant role in governing the dispersed phase size. When two rubbers are generally mixed, if the insignificant phase has lower viscosity as compared to the major phase, then the dispersed phase will be a minor phase. The scattered phase morphology is intensely affected by the composition of the blends. If the major phase shows lower viscosity as compared to the minor phase, an uneven isolated phase will be established (Harrats, 2009). A viscosity ratio near unity produces the smallest size of a particle, and adequate dispersion is attained. As the viscosity ratio moves from unity in any direction, the isolated particles normally become larger. In the circumstance of blends having equimolar components, a component having lower viscosity struggles to compress the component with higher viscosity. Improved dispersion is accomplished when the viscosity ratio of both the phases is near to one another. If the viscosity of the minor phase is high, then the phase can't get fracture and is broken down into very small dispersed particles (Sirisinha et al., 2001; Thomas et al., 2006). The compound having high viscosity can be dispersed better if it is pre-mixed with the plasticizer to bring the viscosity nearer to that of a component with low viscosity. The ratio of blend viscosity disturbs the connectivity of phases and the phase reversal region forming the co-continuous phase morphology. The region of phase inversion becomes smaller with the increase in interfacial tension and is moved along the axis of concentration by changing the ratio of the viscosity of the blend restrictions. In the blending of diverse rubbers, the aim is to accomplish the phase isolated blend that syndicates the properties of original rubbers (Visakh et al., 2013). The size of dispersed polymer blends are also dependent on the polarity amongst the phases as compared to the ratio of viscosity. When the two rubber phases are matched nearly concerning the polarity, the growth of the phase is rapid. This is correct as the viscosity of polymer components approaches one another. The polarity difference of rubbers leads to high interfacial tension and will strictly restrict mixing at an interface and hence the chance for cross-linking amongst the rubbers. It also leads to substandard phase morphology which normally is categorized by the large phase size. Grounded on the comparative polarity, the blends of rubber can be heterogeneous miscible blends or homogeneous miscible blends (Hassander, 1985).

3.12. MORPHOLOGY CATEGORIZATION BY MICROSCOPY

Microscopic imaging in the field of materials science provides a pictorial demonstration of the data and is beneficial in extracting critical parameters. Morphological analysis of the blends of rubber utilizing microscopy is very significant. In light and electron microscopic exploration, the miscibility of blends of the rubber depends on how a particular blend is observed. The image attained from this analysis can be utilized for the abstraction of morphological data and parameters. The blend can normally be homogeneous or miscible if the observation space scale is higher as compared to the size of the domain of diverse phases in the rubber blend.

Blends of natural and synthetic rubbers reveal the morphological hierarchy and multifaceted structure, which govern their performance. Therefore, research goals to visualize the morphologies with the help of microscopy techniques. Several well-developed techniques like electron microscopy (TEM; SEM) and OM techniques allow observing structures of the rubbers and blends at diverse magnification levels. Electron and the scanning probe microscopy techniques can perceive the morphological details at the length scale from visible range to a few nm ranges.

3.13. PREPARATION METHODS UTILIZED FOR MICROSCOPY

The OM samples have been made utilizing cryotome. The sections of one-mm thickness will be kept over the visual micro slides and placed under an optical microscope for assessment (Figure 3.8). The samples can be made by scratching the surface with the help of a razor blade (Liu et al., 2018).

In SEM analysis, the samples were geared up by fracturing under the atmosphere of liquid nitrogen, and the ruptured surface is kept on a stub and then vapor-coated/sputter-coated/utilizing the thin layer of a conducting material such as C, Au, Au/Pd, graphite, and Pt (Lipińska and Imiela, 2019). After coating, the sample is placed inside the chamber of sample for SEM analysis (Figure 3.8). Samples can be made by ebonite approach, in which the samples of rubber are immersed in the molten mixture of sulfenamide accelerator, zinc stearate, and sulfur in a 5:5:90 ratio for around 8 hours at 120°C. Thin samples were normally cut from an immersed sample after eliminating the additional sulfur (S) from the surface (Mathai and Thomas, 2005).

The sample making for TEM analysis is a time-consuming and tedious process. Cryo-ultramicrotomy is a technique utilized for the preparation of a sample for TEM. This technique is primarily utilized for softer composites and polymers. In this technique, samples are normally cooled below room temperature and then sliced using the particular cutter to obtain ultra-thin or thin sections. The specimens of rubber are cooled below Tg (glass transition temperature) and microtomed with the cooled diamond knife. The sections of cryotome were then gathered on the copper grid and placed on a TEM sample holder (as displayed in Figure 3.8).

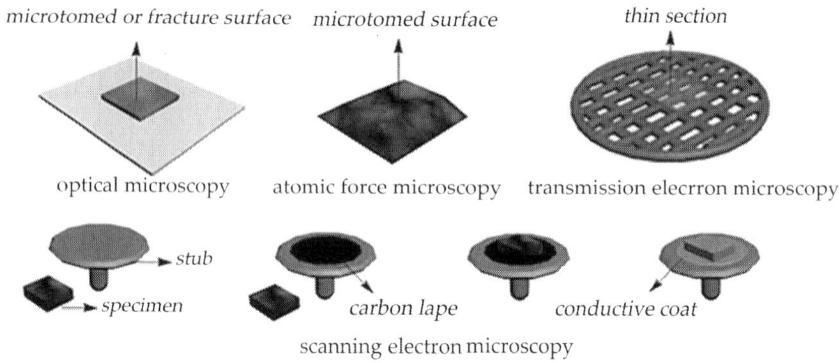

Figure 3.8. Preparation of sample for microscopy analysis.

Source: https://www.researchgate.net/figure/Illustration-of-specimen-preparation-N-8-for-each-analytical-step_fig1_336173339.

For AFM analysis, flat surfaces independent from defects and pollution are needed. Normally, in the situation of the blends of rubber, cryo-microtomed block faces are utilized for analysis (Figure 3.8). The manufactures of microtomes have introduced an exceptional AFM holder in order to prepare samples free from any defects. These holders permit the making of block face and the direct investigation in an AFM without transporting the sample from an insert.

The contrast of diverse phases of the polymer samples is frequently poor as the rubbers are primarily made up of similar elements of light.

Phase-contrast improvements are to be well-thought-out to distinguish rubber components in the blends of rubber, as most of the rubber compounds comprise the large volume of CB; they are typically categorized by electron microscopy. Etching, swelling, staining, and freezing are the techniques utilized to enhance the contrast amongst different phases in the blend of rubber. Marsch et al. (1967) examined immersion of the blend of rubber in a particular solvent such that one component swells more as compared to the other components, and the swollen sample was overextended and then the solvent was evaporated, further, the swollen phase after solvent evaporation gets more thinned, providing the essential contrast. The differential pyrolysis is utilized to selectively eliminate one domain much more transmissive as compared to the domains of stable components. Hess and the co-workers made thin samples of NR/BR rubber blends and then mounted them on the copper grid. The samples were heated under the vacuum to a temperature of nearly 350°C. The NR degrades considerably at a lower temperature as compared to BR and is removed selectively. BR zones seem darker compared to the etched regions representing NR (Hess et al., 1985).

In the ebonite method, one rubber phase is cross-linked and then hardened utilizing the mixture of sulfenamide, zinc stearate, and sulfur, increasing the electron density of one phase, making it simple to recognize phases during the electron microscopy imaging. The higher zinc and sulfur content in a harder phase makes the phase darker as compared to the nearby softer phase because of more electron smattering by zinc and sulfur atoms. The harder phase is much more stable in an electron beam and offers greater contrast amongst the phases. The ebonite approach has the benefit of precise geometrical association within the sample, and the uncured un-compounded rubbers can be utilized as the preliminary materials. The cryogenic method is where the sample is frozen below Tg (glass transition temperature) before microtome, to differentiate between phases in the rubber blends (Smith and Andries, 1974).

Contrast improvement utilizing chemical discoloration agents has been broadly practiced in the research area of rubber. The major reason for the chemical staining is because of a small difference in the electron densities in the rubber blend structure. The chemical staining tempts contrast by placing heavy compounds of metal into one or more than one component of samples under examination. Currently, the main chemical staining agents used are OsO_4 (osmium tetroxide) and RuO_4 (ruthenium tetroxide). Staining by osmium tetroxide has been widely used to enhance phase contrast for the blends of rubber comprising unsaturated rubber as one component, as it stains rubbers with the double bond (Smith and Bryg, 2006). Andrews and Stubbs presented the staining and hardening of the rubber phases with osmium tetroxide. Ruthenium tetroxide diffuses favorably into the amorphous section and then tinges them, and the crystalline section becomes unchanged. Staining is normally performed utilizing vapors of RuO_4 and

OsO_4 solutions. The effectiveness of staining is dependent on the material and the concentration of the staining agent. Staining of the sections of polymer with RuO_4 vapor occurs quickly. OsO_4 takes some minutes to hours or days (Hassander, 1985).

AFM concludes phases in the blends of rubber through stiffness measurements along with imaging. SEM portrays surface morphology; TEM inspects morphological characteristics in bulk in the rubber blends through imaging. These classical methods and the applications to blends of the rubber and composites are given below (Chiu et al., 2006; Khalaf et al., 2018).

3.14. OM (OPTICAL MICROSCOPY) OF THE RUBBER BLENDS

The first examination of the blends of rubber can be carried out by visual assessment. Better comprehensive scrutiny of the blends of rubber can be done utilizing OM. OM utilizes lenses and visible light to acquire magnified images of the rubber blends, therefore also is known as light microscopy. OM was the ancient design of a microscope developed in the 17th century. A compound microscope and a simple microscope are two kinds of microscopes accessible. The quality of an optical image is dependent on three different factors optical aberration, contrast, and resolution. Resolution is well-defined as the ability to distinguish two very closely positioned points in objects as separate entities in an image. Resolution is better when the distance splitting the two small points is very small. Optical aberration can normally be well-defined as the collaboration of light with a glass in the lens, making the wave of light to bend or refract differently in diverse regions of the lens. This can be minimized with the help of using numerous lenses. Contrast can be well-defined as differentiation of the structural components in the blend of rubber by different levels of intensity and shall be enhanced by darkfield, differential interference contrast, cross-polarized light, and phase contrast. OM gives information on the length scale of 10 mm. Microscopes utilized for the analysis of rubber blend are the phase-contrast microscope and polarized light microscope (Subramanian and Mehra, 1987).

In the circumstance of blends of the elastomer, there has been little contrast amongst different phases in the rubber blends when perceived with the help of common light microscopy. There isn't any beneficial data that can be acquired beyond the magnification of 1000. Because of the contrast difference attained in the common light microscopy, an extraordinary type of OM has been utilized, which is called phase-contrast OM which normally utilizes the differences in refractive indices amongst one or more than one phase. Blending rubbers provide a material having refractive indices and which frequently lead to the opaque material. The last requirement of the rubber sample in order to be measured under the optical microscope is a very large difference in the refractive indices amongst the lower concentration of filler and rubber blend phases (Jones and Tinker, 1997).

Using OM, the continuous and dispersed phase in the blends of rubber can be acknowledged. Generally, the rubber's lower concentration becomes dispersed and the higher concentration of rubber becomes the continuous phase. At the equal concentration of 2 rubbers, the blend exhibits the co-continuous morphology. Varghese et al. research on optical images of the Ethylene-vinyl acetate (NBR/EVA) blends. The rubber's lower concentration becomes a dispersed phase in the NBR/EVA blends. In the blend of 50 NBR/50 EVA, the morphology appears to be co-continuous. With the increase of concentration of the NBR both of the phases then become continuous (Varghese et al., 1995).

The novel phase contrast microscopes were utilized to assess the nanofiller's dispersion in a rubber blend matrix. Le et al. worked on macro-dispersion of SBR/NR blends filled with ionic liquid (IL)-modified and unmodified carbon nanotubes (CNTs) fillers utilizing an optical microscope. The optical images (Figure 3.9) of the blends of rubber without IL (Figure 3.9(b)) exhibit uneven dispersion of CNTs, creating agglomerates with an average diameter of nearly 40 mm. IL-modified CNT exhibits even distribution of CNT which is specified by the vanishing of CNT agglomerates (Figure 3.9(a)). The macro-dispersion was estimated utilizing the ratio of non-dispersed agglomerates surface to that of an image, A/A0. The macro-dispersion of CNT in the blend of rubber blend decreases to 1.6% from 12% after modification of CNT with IL. Hence Le et al. determined that macrodispersion of all of the rubber blends with IL-modified CNT exhibits improved filler distribution as compared to rubber blend with unchanged CNT (Le et al., 2016).

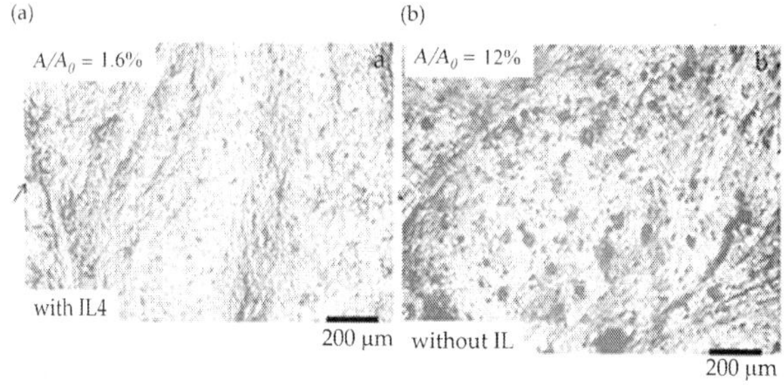

Figure 3.9. Optical images of natural rubber/styrene-butadiene rubber blends with (a) ionic liquid; and (b) without ionic liquid.

Source: https://daneshyari.com/article/preview/5178871.pdf.

Nie et al. examined the morphology of the solution SBR (SSBR)/trans polyisoprene (TPI) rubber blends. They researched phase contrast and the polarized OM with the heating stage from ambient temperature and concluded the experiment at 80°C and observed the separation of phase in SSBR/TPI rubber blends. The phase-separated structure exhibits the bi-continuous morphology, trailed by a progressively roughening and rising process when the blends of SSBR/TPI were heated in equal amounts from room temperature to nearly 80°C and then hardened. At the early stages of analysis, very few variations were observed, as displayed in Figure 3.10(a). The bi-continuous nature gradually busts and the micro-sized swell phase appears (Figure 3.10(b) and (c)). After 140 hours, the sharp domain interface appeared with practically a few droplets of SSBR engulfed into the TPI-loaded domains (Figure 3.10(d)). To guarantee SSBR droplets in the TPI phase aren't the results that seem different at diverse depths, the isolated morphology at diverse focal planes was noticed and confirmed the existence of droplets (Nie et al., 2017).

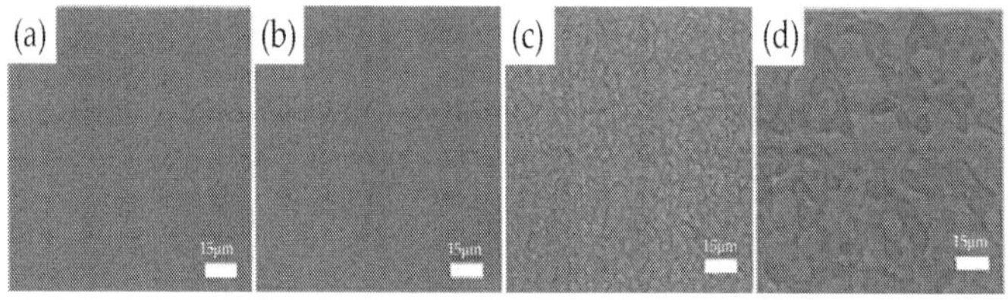

Figure 3.10. Phase morphology of the 50 SSBR/50 trans polyisoprene binary blends hardened at 80°C for (a) 0 hours, (b) 20 hours, (c) 60 hours, and (d) 140 hours.

Source: https://www.sciencedirect.com/science/article/pii/S0032386117303713.

The surface of the ozone-treated specimens of neat silica and rubber blends, rice husk ash, CB-filled rubber blends of NR/EPDM were examined by Arayapranee and Rempel (2008) on the optical microscope. Horizontal lines and fine cracks were observable on surfaces of the unfilled NR/EPDM rubber blends. The surface of the blend of rubber filled with the ash of rice husk appeared to have the largest number of cracks, lower than the blends of unfilled rubber, which can be because of the low adhesion energy between the rubber blends and rice husk ash and uneven dissemination of the ash of rice husk into the rubber blend. The surface cracks also seemed on CB-filled and silica-filled NR/EPDM blends of rubber but were generally less in number than the rice husk ash-filled and unfilled rubber blends, which established that dispersion of fillers and uniform distribution decreased surface cracks of the rubber blends and determines that the reinforcing fillers have improved ozone resistance than their counterparts (Hussein et al., 2004).

Wu and Chen (2011) developed Styrene-butadiene rubber/butadiene rubber blend and the silica-filled SBR/butadiene rubber blends for the optical imaging and established the growth of spherulite in the rubber blends, having an average diameter of nearly 200 mm-even though spherulite in SBR/butadiene rubber blends is problematic owing to the point that spherulite eject and impact one another, while in the silica-filled blends of rubber, no spherulite was noticed, only the crystal string was observable; therefore, Wu and Chen established that blend of SBR/butadiene rubber can't disturb crystal nucleation in the filled blends of rubber but can vary the way of the growth of the crystal, leading to enhancement in properties as the compatibility between BR and SBR upsurges in terms of the crystal morphology (Arayapranee and Rempel, 2007).

Hamed inspected the influence of mixing time on morphology of the EPDM/butadiene rubber blends. He witnessed long parts which were visible with the optical microscope having the mixing time of almost two minutes in the Brabender mixer with five rpm/min, 45°C. Later, he detected that as the time of mixing upsurges long strands started vanishing and the small spherical fragments were observable within the EPDM/butadiene rubber blends (Hamed, 1982).

3.14.1. Cons, Pros, and Artifacts in Optical Microscopy

Morphological examination using OM is very simple and there isn't any extensive preparation of the sample. The resolution of OM is very low around <0.25 m, and no info on the surface can be acquired beyond 1000. Because of this less low resolution, OM is only appropriate for strong and dark refracting materials. In the situation of blends of the rubber, details of morphology or phase structure can't be attained from OM. OM has less resolution as compared to the other microscopy methods. Because of this, magnification with less resolution at higher levels usually makes the image unclear and blur for identification (Canetta et al., 2014). The least level of magnification utilized in measuring the images from an optical microscope varies from 500–1500. While making the thin segments for OM there can normally be cutting artifacts as in the situation of highly loaded blends of rubber. Because of this, dark patterns will be visible in the OM images and they must be removed with the help of image analysis software. The main artifacts in OM contain the deformation of a specimen which can cause variations in the shape of a sample or yield stress-tempted structures that normally can be interpreted just like in the polymer (Conzatti and Galimberti, 2013).

3.15. MORPHOLOGY OF THE RUBBER BLENDS THROUGH SEM

SEM, the electron microscopy method, utilizes the focused beam of electrons in order to produce images of the surface of a sample through point scanning of the surface. Information obtained in the shape of images is because of the collaboration of electrons with the atoms in a sample and provides details associated with samples composition and surface topography (Varkey et al., 2000; Dubey et al., 2006).

SEM is among the most broadly utilized techniques in the early days along with modern time to conclude phase separation in the rubber blends (Eaton and West, 2010).

3.15.1. Scanning Electron Microscopic Examination of the Unfilled Rubber Blends

Morphology of the blends of rubber is primarily 1 component dispersed in a continuous matrix of other or both of the phases become continuous all over the blend and is called the co-continuous morphology. Generally, the higher fraction of components becomes a continuous matrix and the lower component fraction becomes the dispersed phase. In the binary blend of A and B, the morphology of phase can be of the B droplet in an A matrix when generally B is minor. The component which is minor will be dispersed in a matrix if the viscosity is lower as compared to that of a matrix. Figure 3.11 signifies the dispersed morphology of phase in the rubber blends (Smith, 1967; Go and Ha, 1996). The morphology of the surface of the rubber blends was concluded utilizing SEM analysis. In unfilled blends, the continuous and dispersed phase can be recognized with the help of SEM images. With the upsurge of the content of one rubber the domain size upsurges. With the increase of the content of one of the phases, the rubber with lower concentration becomes the dispersed phase. The co-continuous phase gives the morphology where 2 phases co-exist with one another and the blend shows the superlative performance. With the increase of concentration of the second rubber, a continuous phase is formed and phase inversion takes place (Koshy et al., 1992; Koseki et al., 1999).

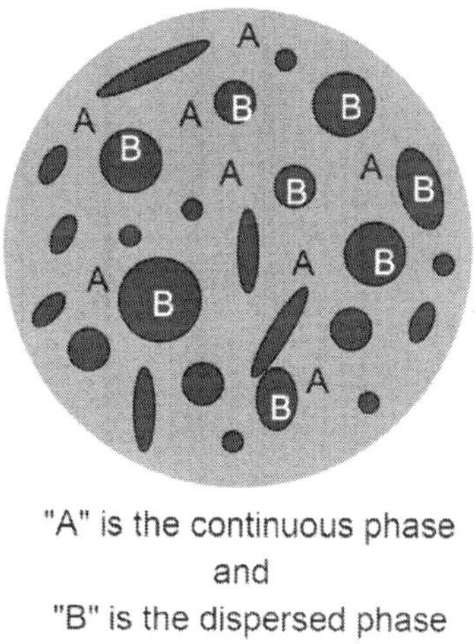

Figure 3.11. Schematic representation of continuous and dispersed phases in the rubber-rubber blends.

Source: https://www.x-mol.com/paper/1278031921090285568.

Varghese et al. (1995) anticipated diverse morphology forms in the blends of NBR/EVA as the ratio of the components differs. At the lower concentration of NBR, NBR creates the dispersed phase and EVA behaves as the continuous phase. In a 50/50 ratio of the blend, both of the phases generally become continuous. The size of a phase of the dispersed domains of EVA and NBR increases with an upsurge in NBR and EVA contents, respectively (Kim et al., 2006; Shan et al., 2011). The main reason for the upsurge in the size of the domain is the recombination or coalescence behavior of the domains. For 60/40 NR/EVA blends, a thin layer created is broken because of its incapability to survive the forces because of the swelling of the NR phase. Finally, images of SEM validate that the EVA forms a dispersed phase when the content of EVA equals 40%, and NR forms a dispersed phase at 60% or a fewer amount (Ricci and Braga, 2004; Smith and Bryg, 2006). In a study, Sidkey, and the co-workers researched on polyurethane (PU)/NR blends, NR seems as bright and dark regions whereas PU seems to be elongated and large lines creating opaque domains (Sidkey et al., 2002; Microscopy, 2013).

3.15.2. SEM Analysis of the Compatibilized Unfilled Blends of Rubber

Compatibilizers are added to upsurge the interaction amongst the phases of rubber. The upsurged interaction amongst phases produces more of the tear lines and seems on the images of SEM of the fracture surface. The tear line shows the enhanced interaction amongst the phases and enhanced compatibility which outcomes in enhanced tensile strength. The uncompatibilized blends display rough and coarser morphology than the compatibilized blends (Razavi-Nouri et al., 2017; Gatos and Karger-Kocsis, 2010). Noriman et al. (2010) demonstrated SEM micrographs for ductile fracture surfaces of the SBR/recycled NBR blends, compatibilized with ENR-50 (epoxidized NR-50) and SBR/recycled-NBR blends without epoxidized NR-50 at the ratio of nearly 95/5. Numerous tear lines with the rougher surface infer additional energy required to rupture the sample than the wider tearing line. This validates the superior interface amongst the SBR and r-NBR matrix with the adding of epoxidized NR-50 as the compatibilizer (Monfared and Jalali-Arani, 2015). SEM images of the failure surface 95 SBR/5 recycled-NBR exhibit a rough surface with several tearing lines having branching, representing vigorous interfacial adhesion with recycled-NBR as compared to virgin NBR. In a similar kind of effort by Noriman et al. (2012) centered on SBR/recycled-NBR, SEM images of the ductile-fractured surface of the SBR/recycled-NBR blends at the 50/50 ratio of the blend were witnessed (George et al., 2000; Shan et al., 2011).

The incompatible NR/CSM (chlorosulfonated polyethylene) blends were compatibilized utilizing ENR-20, PVC (polyvinylchloride), ENR-50, and Struktol. Integration of compatibilizers enhances interfacial adhesion amongst the dispersed NR particles in the chlorosulfonated polyethylene (CSPE) matrix. After the adding of a blend of ENR-20, compatibility is enhanced, and this is displayed by the reduced number of particle cavities. Epoxidized NR-50 intermingles with the CSM in blends, which might root a reduction in the interfacial stress between CSM and NR, ensuing in an enhancement in the compatibility of NR/CSM blends (Kim et al., 2006). The ternary blends of CSM/ENR/NR at 30/5/65 didn't show any variation in morphology at 2000 X but at 10,000 X 1-micrometer-sized particle size of the CSM is noticed. While at 60/10/30 domains of the submicron size of the CSM particles are noticed. No variation in morphology for the 70/30 blends of rubber was witnessed after the addition of the Struktol. Suma et al. (1993) inspected fractured samples of EPDM/NR and NR/IIR (isobutylene-isoprene) blends of rubber. The cracked surfaces of the improved blend displayed a smoother and homogeneous pattern (Smith and Bryg, 2006; Boonsong et al., 2008).

Microphotographs attained from the SEM of the cryo-fractured surfaces for 30 FKM (fluoroelastomer)/70 NR blend with and without the third component are exhibited by Phiriyawirut et al. (2013). Existence of holes in a matrix can be well-thought-out as an existence of the phase separation happening during the process of fracture, representing insufficient adhesion amongst blending components. Improvement incompatibility of the blend was carried out by adding the third component with the segment of PIME or ENT, which was identical chemically to FKM and NR phases. Roughness in the surface of the blend of rubber can be noticed after adding ENR, but a distinguished effect on the morphology of the NR/FKM blend wasn't observed (Kim et al., 2006; Phiriyawirut and Luamlam, 2013).

3.15.3. SEM of Diverse Curative Filled Blends of Rubber

From the images of SEM of the rubber blends, the upsurged loading of the curative agents upsurges the roughness of the surface of the fractured specimens because of branching. The steps of vulcanization impact the morphology of the blend. In 1 and 2-step vulcanization, numerous vacuoles were noticed in the previous, and coherency is perceived in the latter. 2-stage vulcanization decreases the number of vacuoles and it develops compact morphology in the rubber blend (Phiriyawirut et al., 2013; Visconte et al., 2004).

3.15.4. Scanning Electron Microscopic Examination of the Filled Rubber Blends

SEM images of the filled rubber blend have a rough surface, whereas unfilled blends display a smooth surface. The coarse surface is because of stiff phases that ascend from the improved interaction between the filler particles and elastomer. Azizli et al. (2017) researched the impact of cloisite clay on the morphology of the IIR rubber/EPDM compounds. The SEM pictures of cryo-fractured surfaces display some rough and smooth spots. They have discovered that the inclusion of one phr of nano clay disturbs the fracture behavior and morphology (Gu et al., 2009; Mohan et al., 2011).

Prasertsri et al. (2013) developed NR/SBR blends and researched the impact of upsurging silica loading. It was perceived that at lower silica loading in NR/SBR blends, dispersion and distribution took place efficiently, whereas on increasing the silica concentration, large aggregates are observable in the SEM images. The impact of 30 phr silica on the NR/SBR blends at changing concentration was witnessed. NR-rich blends had good distribution and dispersion of silica-whereas for SBR-ample phases aggregates of the few micrometers were noticed (Kantala et al., 2009; Raji et al., 2015).

Tavakoli et al. (2012) invented an organoclay-filled NR/SBR blend with EPDM grafted maleic anhydride (EPDM-g-MAH) as a compatibilizer. Cryo-fractured specimens of NR/SBR blends were inspected in the existence and nonexistence of the compatibilizer. The uncompatibilized rubber blends displayed a smooth fractured surface representing a weak interface amongst nanoclay and the polymer matrix. The compatibilized rubber blends displayed high surface roughness representing an improved interface between the polymer matrix and the nanoclay (Arroyo et al., 2007).

Zarei et al. (2017) made a ternary blend of NR/SBR and BR without and with nano clay as addition and characterized for the morphological examination. Zarei et al. developed ternary nanocomposites with 1, 3, 5, 7, and 10 wt.% to have a comprehensive understanding of the impact of nano clay on the ternary rubber matrix. It was observed that the coarse surface on the inclusion of nano clay was because of good interaction amongst the clay and ternary blend matrix. Conversely, the ternary blend matrix without the nano clay was distinguished from the nano clay-filled ternary blend because of the smooth surface than the filled ternary blend (Andideh et al., 2014).

Micro silica dispersion and nano-silica dispersion in NBR/CSM and NBR/SBR were assessed by Markovi et al. (2004); and Perez et al. (2008)

Silica dispersion upsurges with the increase of the content of the NBR because of the higher attraction of silica towards the polar rubber phase. The micro-filled rubber blends have a rough surface, and the nano-filled blend exhibits a more polished structure. The inclusion of the compatibilizer (triethoxy silyl propyl tetrasulfide (TESPT)) enhances the silica dispersion in the rubber blends (Pal et al., 2010; Perez and Lopez, 2012).

REFERENCES

1. Andideh, M., Naderi, G., Ghoreishy, M. H. R., & Soltani, S., (2014). Effects of nanoclay and short nylon fiber on morphology and mechanical properties of nanocomposites based on NR/SBR. *Fibers and Polymers, 15*(4), 814–822.
2. Arayapranee, W., & Rempel, G. L., (2007). Properties of NR/EPDM blends with or without methyl methacrylate-butadiene-styrene (MBS) as a compatibilizer. *International Journal of Materials and Structural Reliability, 5*(1), 1–12.
3. Arayapranee, W., & Rempel, G. L., (2008). Morphology and mechanical properties of natural rubber and styrene-grafted natural rubber latex compounds. *Journal of Applied Polymer Science, 109*(3), 1395–1402.
4. Arroyo, M., Lopez-Manchado, M. A., Valentin, J. L., & Carretero, J., (2007). Morphology/ behavior relationship of nanocomposites based on natural rubber/epoxidized natural rubber blends. *Composites Science and Technology, 67*(7, 8), 1330–1339.
5. Asakura, T., & Ando, I., (1998). *Solid-state NMR of polymers* (Vol. 1, pp. 1–33). Elsevier.
6. Azizli, M. J., Abbasizadeh, S., Hoseini, M., Rezaeinia, S., & Azizli, E., (2017). Influence of blend composition and organic cloisite 15A content in the structure of isobutylene-isoprene rubber/ethylene propylene diene monomer composites for investigation of morphology and mechanical properties. *Journal of Composite Materials, 51*(13), 1861–1873.a2qqqqq.
7. Bischoff, P. H., & Scanlon, A., (2007). Effective moment of inertia for calculating deflections of concrete members containing steel reinforcement and fiber-reinforced polymer reinforcement. *ACI Structural Journal, 104*(1), 68.
8. Boonsong, K., Seadan, M., & Lopattananon, N., (2008). Compatibilization of natural rubber (NR) and chlorosulfonated polyethylene (CSM) blends with zinc salts of sulfonated natural rubber. *Songklanakarin Journal of Science and Technology, 30*(4), 1–22.
9. Botros, S. H., (2002). Preparation and characteristics of NR/EPDM rubber blends. *Polymer-Plastics Technology and Engineering, 41*(2), 341–359.

10. Botros, S. H., Moustafa, A. F., & Ibrahim, S. A., (2006). Homogeneous styrene-butadiene/acrylonitrile butadiene rubber blends. *Polymer-Plastics Technology and Engineering, 45*(4), 503–512.
11. Canetta, E., Riches, A., Borger, E., Herrington, S., Dholakia, K., & Adya, A. K., (2014). Discrimination of bladder cancer cells from normal urothelial cells with high specificity and sensitivity: Combined application of atomic force microscopy and modulated Raman spectroscopy. *Acta Biomaterialia, 10*(5), 2043–2055.
12. Cappella, B., Geuss, M., Klüppel, M., Munz, M., Schulz, E., & Sturm, H., (2004). *Filler-Reinforced Elastomers Scanning Force Microscopy* (Vol. 164, pp. 1–33). Springer.
13. Chang, Y. W., Shin, Y. S., Chun, H., & Nah, C., (1999). Effects of trans-polyoctylene rubber (TOR) on the properties of NR/EPDM blends. *Journal of Applied Polymer Science, 73*(5), 749–756.
14. Chiu, H. T., Li, W. C., & Tsai, P. A., (2006). A study of rheological behavior and compatibility of NR/BR blends. *Polymer-Plastics Technology and Engineering, 45*(7), 845–855.
15. Conzatti, L., & Galimberti, M., (2013). Microscopy of natural rubber composites and nanocomposites. In: *Natural Rubber Materials* (Vol. 1, pp. 649–682).
16. Corish, P. J., (1967). Fundamental studies of rubber blends. *Rubber Chemistry and Technology, 40*(2), 324–340.
17. Das, A., Stöckelhuber, K. W., Jurk, R., Saphiannikova, M., Fritzsche, J., Lorenz, H., & Heinrich, G., (2008). Modified and unmodified multiwalled carbon nanotubes in high performance solution-styrene-butadiene and butadiene rubber blends. *Polymer, 49*(24), 5276–5283.
18. De Luzuriaga, A., (2008). Key role of entropy in nanoparticle dispersion: Polystyrene-nanoparticle/linear-polystyrene nanocomposites as a model system. *Physical Chemistry Chemical Physics, 10*(5), 650–651.
19. De, S. K., & White, J. R., (2001). *Rubber Technologist's Handbook* (Vol. 1, pp. 1–22). Smithers Rapra Publishing.
20. Donnet, J. B., & Custodero, E., (2005). Reinforcement of elastomers by particulate fillers. In: *Science and Technology of Rubber* (Vol. 1, pp. 367–400). Academic Press.
21. Dubey, K. A., Bhardwaj, Y. K., Chaudhari, C. V., & Sabharwal, S., (2006). Radiation-processed styrene-butadiene-co-ethylene-propylene

diene rubber blends: Compatibility and swelling studies. *Journal of Applied Polymer Science, 99*(6), 3638–3649.

22. Eaton, P., & West, P., (2010). *Atomic Force Microscopy* (Vol. 1, pp. 1–33). Oxford University Press.
23. El-Sabbagh, S. H., (2003). Compatibility study of natural rubber and ethylene-propylene-diene rubber blends. *Polymer Testing, 22*(1), 93–100.
24. Fenouillot, F., Cassagnau, P., & Majesté, J. C., (2009). Uneven distribution of nanoparticles in immiscible fluids: Morphology development in polymer blends. *Polymer, 50*(6), 1333–1350.
25. Frisch, H. L., Klempner, D., & Frisch, K. C., (1970). A Topologically Interpenetrating Elastomeric Network. *Rubber Chemistry and Technology, 43*(4), 883–886.
26. Fujimaki, T., & Oshima, N., (1989). *U.S. Patent No. 4,866,131* (Vol. 1, pp. 1–33). Washington, DC: U.S. Patent and Trademark Office.
27. Fujimoto, K., & Yoshimiya, N., (1968). Blends of cis-1,4-polybutadiene with natural or styrene-butadiene rubber. *Rubber Chemistry and Technology, 41*(3), 669–677.
28. Gatos, K. G., & Karger-Kocsis, J., (2010). Rubber/clay nanocomposites: Preparation, properties and applications. *Rubber Nanocomposites: Preparation, Properties, and Applications, 1*, 169–195.
29. George, S. C., Ninan, K. N., Groeninckx, G., & Thomas, S., (2000). Styrene-butadiene rubber/natural rubber blends: Morphology, transport behavior, and dynamic mechanical and mechanical properties. *Journal of Applied Polymer Science, 78*(6), 1280–1303.
30. Ghilarducci, A., Cerveny, S., Salva, H., Matteo, C. L., & Marzocca, A. J., (2001). Influence of the blend composition in the internal friction of NR/SBR compounds. *Kautschuk Gummi Kunststoffe, 54*(7, 8), 382–386.
31. Ginzburg, V. V., (2010). Nanoparticle/polymer blends: Theory and modeling. *Encyclopedia of Polymer Blends, 1*, 233–268.
32. Go, J. H., & Ha, C. S., (1996). Rheology and properties of EPDM/BR blends with or without a homogenizing agent or a coupling agent. *Journal of Applied Polymer Science, 62*(3), 509–521.
33. Go, J. H., Ha, C. S., & Noh, I., (1995). Effects of a compatibilizer on the properties of EPDM/BR blends. *Korea Polymer Journal, 3*(1), 25–34.

34. Gooch, J. W., (2010). *Encyclopedic Dictionary of Polymers* (Vol. 1, pp. 1–33). Springer Science and Business Media.

35. Gu, Z., Song, G., Liu, W., Li, P., Gao, L., Li, H., & Hu, X., (2009). Preparation and properties of styrene-butadiene rubber/natural rubber/organo-bentonite nanocomposites prepared from latex dispersions. *Applied Clay Science, 46*(3), 241–244.

36. Hamed, G. R., (1982). Morphology-property relationships in EPDM-polybutadiene blends. *Rubber Chemistry and Technology, 55*(1), 151–160.

37. Hansen, C. M. (1967). The three dimensional solubility parameter and solvent diffusion coefficient: Their importance in surface coating formulation, 1, 1-33.

38. Harrats, C., (2009). *Multiphase Polymer-Based Materials: An Atlas of Phase Morphology at the Nano and Micro Scale* (Vol. 1, pp. 1–22). CRC press.

39. Hassander, H., (1985). Electron microscopy methods for studying polymer blends—comparison of scanning electron microscopy and transmission electron microscopy. *Polymer Testing, 5*(1), 27–36.

40. Hess, W. M., Vegvari, P. C., & Swor, R. A., (1985). Carbon black in NR/BR blends for truck tires. *Rubber Chemistry and Technology, 58*(2), 350–382.

41. Hussein, I. A., Chaudhry, R. A., & Abu, S. B. F., (2004). Study of the miscibility and mechanical properties of NBR/HNBR blends. *Polymer Engineering and Science, 44*(12), 2346–2352.

42. Ifuku, S., Nogi, M., Abe, K., Yoshioka, M., Morimoto, M., Saimoto, H., & Yano, H., (2011). Simple preparation method of chitin nanofibers with a uniform width of 10–20 nm from prawn shell under neutral conditions. *Carbohydrate Polymers, 84*(2), 762–764.

43. Inoue, T., Shomura, F., Ougizawa, T., & Miyasaka, K., (1985). Co-vulcanization of polymer blends. *Rubber Chemistry and Technology, 58*(5), 873–884.

44. Isayev, A. I., & Palsule, S., (2011). *Encyclopedia of Polymer Blends, Volume 2: Processing,* 1–33. John Wiley & Sons.

45. Jones, K. C., & Tinker, A., (1997). *Blends of Natural Rubber: Novel Techniques for Blending with Specialty Polymers* (Vol. 1, pp. 1–22). Springer Science and Business Media.

46. Jose, L., & Joseph, R., (2005). Latex stage blending of natural rubber and poly (vinyl chloride) for improved mechanical properties. *International Journal of Polymeric Materials, 54*(5), 387–396.
47. Joseph, R., George, K. E., & Francis, D. J., (1988). Scanning electron microscopy examination of the fracture surface of NR/SBR and NR/BR blends. *International Journal of Polymeric Materials, 12*(2), 111–123.
48. Kantala, C., Wimolmala, E., Sirisinha, C., & Sombatsompop, N., (2009). Reinforcement of compatibilized NR/NBR blends by fly ash particles and precipitated silica. *Polymers for Advanced Technologies, 20*(5), 448–458.
49. Khalaf, A. I., Ward, A. A., & Rozik, N. N., (2018). Investigation of physical properties and morphology of compatibilized EPDM/EVA blends. *Journal of Thermoplastic Composite Materials, 31*(3), 376–391.
50. Kim, S. K., Kang, J. H., Choe, Y., & Chang, Y. W., (2006). Structure and properties of the organoclay filled NR/BR nanocomposites. *Macromolecular Research, 14*(2), 187–193.
51. Koseki, Y., Lee, M. S., & Macosko, C. W., (1999). Encapsulation in ternary elastomer blends. *Rubber Chemistry and Technology, 72*(1), 109–118.
52. Koshy, A. T., Kuriakose, B., & Thomas, S., (1992). Studies on the effect of blend ratio and cure system on the degradation of natural rubber—ethylene-vinyl acetate rubber blends. *Polymer Degradation and Stability, 36*(2), 137–147.
53. Le, H. H., Das, A., Basak, S., Tahir, M., Wießner, S., Fischer, D., & Radusch, H. J., (2016). Effect of different ionic liquids on the dispersion and phase selective wetting of carbon nanotubes in rubber blends. *Polymer, 105*, 284–297.
54. Li, M. C., & Cho, U. R., (2017). Starch in rubber based blends and micro composites. In: *Rubber Based Bio-Nanocomposites* (Vol. 1, pp. 109–140). Springer, Cham.
55. Limper, A., (2012). *Mixing of Rubber Compounds* (Vol. 1, pp. 1–33). Carl Hanser Verlag GmbH Co KG.
56. Lipińska, M., & Imiela, M., (2019). Morphology, rheology and curing of (ethylene-propylene elastomer/hydrogenate acrylonitrile-butadiene

rubber) blends reinforced by POSS and organoclay. *Polymer Testing, 75*, 26–37.

57. Liu, X., Bian, L., Gao, Y., & Wang, Z., (2012). Influence of ultrafine full-vulcanized powdered rubber on NR/SBR blends. *Polymer Bulletin, 69*(6), 747–756.
58. Liu, X., Dong, X., Liu, W., Han, C. C., & Wang, D., (2018). Morphology evolution and dynamic relaxation behavior of solution-polymerized styrene-butadiene rubber/polyisoprene/silica ternary composites influenced by shear. *Polymer, 145*, 416–425.
59. Mangaraj, D., (2004). Rubber recycling by blending with plastics. *Rubber Recycling, 1*, 260–272.
60. Maria, H. J., Lyczko, N., Nzihou, A., Joseph, K., Mathew, C., & Thomas, S., (2014). Stress relaxation behavior of organically modified montmorillonite filled natural rubber/nitrile rubber nanocomposites. *Applied Clay Science, 87*, 120–128.
61. Markovic, G., Radovanovic, B., Budinski-Simendić, J. K., & Marinović-Cincović, M., (2004). Thermostability and surface morphology of nano-and micro-filled NBR/CSM and CR/CSM rubber blends. *Journal of the Serbian Chemical Society, 69*(2), 167–173.
62. Marsh, P. A., Voet, A., & Price, L. D., (1967). Electron microscopy of heterogeneous elastomer blends. *Rubber Chemistry and Technology, 40*(2), 359–370.
63. Mathai, A. E., & Thomas, S., (2005). Morphology, mechanical and viscoelastic properties of nitrile rubber/epoxidized natural rubber blends. *Journal of Applied Polymer Science, 97*(4), 1561–1573.
64. Michler, G. H., (2008). *Transmission Electron Microscopy: Conventional and Special Investigations of Polymers, 1*, 1–23.
65. Microscopy, A., (2013). Microanalysis research facility. *MyScope: Training for Advance Research, 1, 1–33.*
66. Mohan, T. P., Kuriakose, J., & Kanny, K., (2011). Effect of nanoclay reinforcement on structure, thermal and mechanical properties of natural rubber-styrene butadiene rubber (NR-SBR). *Journal of Industrial and Engineering Chemistry, 17*(2), 264–270.
67. Mondal, S., & Khastgir, D., (2017). Evaluation of carbon black distribution in different phases of compatible blend of EVA/NBR through electrical, mechanical and morphological test. *Polymer Testing, 59*, 404–413.

68. Monfared, A., & Jalali-Arani, A., (2015). Morphology and rheology of (styrene-butadiene rubber/acrylonitrile-butadiene rubber) blends filled with organoclay: The effect of nanoparticle localization. *Applied Clay Science, 108*, 1–11.

69. Mutz, M., (2016). Thermodynamics, phase segregation, polymer morphology, and interface/interphases in polymer nanocomposites. In: *Spectroscopy of Polymer Nanocomposites* (Vol. 1, pp. 65–87). William Andrew Publishing.

70. Naseri, A. S. Z., & Jalali-Arani, A., (2015). A comparison between the effects of gamma radiation and sulfur cure system on the microstructure and crosslink network of (styrene-butadiene rubber/ethylene propylene diene monomer) blends in the presence of nanoclay. *Radiation Physics and Chemistry, 115*, 68–74.

71. Naskar, N., Biswas, T., & Basu, D. K., (1994). Polymer blend: A novel method for the preparation of a natural rubber-carboxylated nitrile rubber blend. *Journal of Applied Polymer Science, 52*(8), 1007–1014.

72. Nie, H., Liu, D., Liu, C., Wang, X., & He, A., (2017). Morphology evolution in solution polymerized styrene-butadiene rubber (SSBR)/trans-1, 4-polyisoprene (TPI) blends: SSBR particle formation, TPI crystal nucleation, growth and polymorphic form. *Polymer, 117*, 11–16.

73. Noriman, N. Z., Ismail, H., & Rashid, A. A., (2010). Characterization of styrene-butadiene rubber/recycled acrylonitrile-butadiene rubber (SBR/NBRr) blends: The effects of epoxidized natural rubber (ENR-50) as a compatibilizer. *Polymer Testing, 29*(2), 200–208.

74. Noriman, N. Z., Ismail, H., & Rashid, A. A., (2012). Properties of styrene-butadiene rubber/recycled acrylonitrile-butadiene rubber (SBR/NBRr) blends: Effect of the addition of trans-polyoctylene rubber. *Journal of Applied Polymer Science, 126*(S2), E56–E63.

75. Pal, K., Rajasekar, R., Kang, D. J., Zhang, Z. X., Pal, S. K., Das, C. K., & Kim, J. K., (2010). Influence of carbon blacks on butadiene rubber/high styrene rubber/natural rubber with nano-silica: Morphology and wear. *Materials and Design, 31*(3), 1156–1164.

76. Pandey, K. N., Setua, D. K., & Mathur, G. N., (2005). Determination of the compatibility of NBR-EPDM blends by an ultrasonic technique, modulated DSC, dynamic mechanical analysis, and atomic force microscopy. *Polymer Engineering and Science, 45*(9), 1265–1276.

77. Perez, L. D., & Lopez, B. L., (2012). Thermal characterization of SBR/NBR blends reinforced with a mesoporous silica. *Journal of Applied Polymer Science, 125*(S1), E327–E333.
78. Perez, L. D., Sierra, L., & Lopez, B. L., (2008). Effect of the filler characteristics on the miscibility of styrene-butadiene rubber and nitrile butadiene rubber blends. *Polymer Engineering and Science, 48*(10), 1986–1993.
79. Phiriyawirut, M., & Luamlam, S., (2013). Influence of poly (vinyl chloride) on natural rubber/chlorosulfonated polyethylene blends. *Open Journal of Organic Polymer Materials, 2013, 1, 1–34.*
80. Phiriyawirut, M., Limwongwatthananan, T., Kaemram, S., & Wiengkaew, S., (2013). Natural rubber/fluoroelastomer blends: Effect of third component on cure characteristics, morphology, mechanical properties, and automotive fuel swelling. *Open Journal of Polymer Chemistry, 3*(04), 79.
81. Pierson, J., Fernández, J. J., Bos, E., Amini, S., Gnaegi, H., Vos, M., & Peters, P. J., (2010). Improving the technique of vitreous cryo-sectioning for cryo-electron tomography: Electrostatic charging for section attachment and implementation of an anti-contamination glove box. *Journal of Structural Biology, 169*(2), 219–225.
82. Prasertsri, S., Lagarde, F., Rattanasom, N., Sirisinha, C., & Daniel, P., (2013). Raman spectroscopy and thermal analysis of gum and silica-filled NR/SBR blends prepared from latex system. *Polymer Testing, 32*(5), 852–861.
83. Raji, V. V., Ramakrishnan, S., Sukumar, R., Brahmakumar, M., & Menon, A. R., (2015). Kaolin modified with sodium salt of rubber seed oil as a reinforcing filler for blends of natural rubber, polybutadiene rubber and acrylonitrile-butadiene rubber. *Polymer International, 64*(11), 1585–1593.
84. Rathi, A., Hernández, M., Garcia, S. J., Dierkes, W. K., Noordermeer, J. W., Bergmann, C., & Blume, A., (2018). Identifying the effect of aromatic oil on the individual component dynamics of S-SBR/BR blends by broadband dielectric spectroscopy. *Journal of Polymer Science Part B: Polymer Physics, 56*(11), 842–854.
85. Ray, S. S., & Bousmina, M., (2006). Crystallization behavior of Poly [(butylene succinate)-co-adipate] nanocomposite. *Macromolecular Chemistry and Physics, 207*(14), 1207–1219.

86. Razavi-Nouri, M., Karami, M., & Naderi, G., (2017). Rheological behavior and morphology of acrylonitrile-butadiene rubber/poly (ethylene-co-vinyl acetate) blends filled by various organoclay contents. *Applied Clay Science, 145*, 1–10.

87. Ricci, D., & Braga, P. C., (2004). Recognizing and avoiding artifacts in AFM imaging. In: *Atomic Force Microscopy* (Vol. 1, pp. 25–37). Humana Press.

88. Roland, C. M., & Boehm, G. G. A., (1985). Macromolecular diffusion and the auto adhesion of polybutadiene. *Macromolecules, 18*(6), 1310–1314.

89. Saad, A. L., & El-Sabbagh, S., (2001). Compatibility studies on some polymer blend systems by electrical and mechanical techniques. *Journal of Applied Polymer Science, 79*(1), 60–71.

90. Sae-Oui, P., Suchiva, K., Thepsuwan, U., Intiya, W., Yodjun, P., & Sirisinha, C., (2016). Effects of blend ratio and SBR type on properties of silica-filled SBR/NR tire tread compounds. *Rubber Chemistry and Technology, 89*(2), 240–250.

91. Sahakaro, K., Datta, R. N., Baaij, J., & Noordermeer, J. W., (2007). Blending of NR/BR/EPDM by reactive processing for tire sidewall applications. III. Assessment of the blend ozone-and fatigue-resistance in comparison with a conventional NR/BR compound. *Journal of Applied Polymer Science, 103*(4), 2555–2563.

92. Satyanarayana, M. S., Bhowmick, A. K., & Kumar, K. D., (2016). Preferentially fixing nanoclays in the phases of incompatible carboxylated nitrile rubber (XNBR)-natural rubber (NR) blend using thermodynamic approach and its effect on physicomechanical properties. *Polymer, 99*, 21–43.

93. Sawyer, L. C., & Grubb, D. T., (1996). New techniques in polymer microscopy. In: *Polymer Microscopy* (Vol. 1, pp. 315–356). Springer, Dordrecht.

94. Shan, C., Gu, Z., Wang, L., Li, P., Song, G., Gao, Z., & Yang, X., (2011). Preparation, characterization, and application of NR/SBR/Organoclay nanocomposites in the tire industry. *Journal of Applied Polymer Science, 119*(2), 1185–1194.

95. Sidkey, M. A., Yehia, A. A., Abd, E. M. N. A., & Gaafar, M. S., (2002). Compatibility studies on some rubber blend systems by ultrasonic techniques. *Materials Chemistry and Physics, 74*(1), 23–32.

96. Sirisinha, C., Baulek-Limcharoen, S., & Thunyarittikorn, J., (2001). Changes in morphology and properties of NR-NBR blends. Effect of viscosity ratio modified by liquid natural rubber and epoxidized liquid natural rubber. *Plastics, Rubber and Composites, 30*(7), 314–317.
97. Sirqueira, A. S., & Soares, B. G., (2002). Mercapto-modified copolymers in elastomer blends. IV. The compatibilization of natural rubber/EPDM blends. *Journal of Applied Polymer Science, 83*(13), 2892–2900.
98. Smith, R. W., & Andries, J. C., (1974). New methods for electron microscopy of polymer blends. *Rubber Chemistry and Technology, 47*(1), 64–78.
99. Smith, R. W., & Bryg, V., (2006). Staining polymers for microscopical examination. *Rubber Chemistry and Technology, 79*(3), 520–540.
100. Smith, R. W., (1967). Vacuole formation and the Mullins effect in SBR blends with polybutadiene. *Rubber Chemistry and Technology, 40*(2), 350–358.
101. Stephen, R., Ranganathaiah, C., Varghese, S., Joseph, K., & Thomas, S., (2006). Gas transport through nano and micro composites of natural rubber (NR) and their blends with carboxylated styrene-butadiene rubber (XSBR) latex membranes. *Polymer, 47*(3), 858–870.
102. Stephens, H. L., & Bhowmick, A. K., (2001). *Handbook of Elastomers* (Vol. 1, pp. 1–30). Dekker.
103. Subramanian, P. M., & Mehra, V., (1987). Laminar morphology in polymer blends: Structure and properties. *Polymer Engineering and Science, 27*(9), 663–668.
104. Sui, G., Zhong, W. H., Yang, X. P., Yu, Y. H., & Zhao, S. H., (2008). Preparation and properties of natural rubber composites reinforced with pretreated carbon nanotubes. *Polymers for Advanced Technologies, 19*(11), 1543–1549.
105. Suma, N., Joseph, R., & George, K. E., (1993). Improved mechanical properties of NR/EPDM and NR/butyl blends by precuring EPDM and butyl. *Journal of Applied Polymer Science, 49*(3), 549–557.
106. Tavakoli, M., Katbab, A. A., & Nazockdast, H., (2012). NR/SBR/organoclay nanocomposites: Effects of molecular interactions upon the clay microstructure and mechano-dynamic properties. *Journal of Applied Polymer Science, 123*(3), 1853–1864.

107. Thomas, S., Harrats, C., & Groeninckx, G., (2006). *Micro-and Nanostructured Polymer Blends: State of the Art, Challenges, and Future Prospects* (Vol. 1, pp. 1–42). Taylor & Francis Group, LLC: Boca Raton, FL.
108. Varghese, H., Bhagawan, S. S., Rao, S. S., & Thomas, S., (1995). Morphology, mechanical and viscoelastic behavior of blends of nitrile rubber and ethylene-vinyl acetate copolymer. *European Polymer Journal, 31*(10), 957–967.
109. Varkey, J. T., Augustine, S., Groeninckx, G., Bhagawan, S. S., Rao, S. S., & Thomas, S., (2000). Morphology and mechanical and viscoelastic properties of natural rubber and styrene-butadiene rubber latex blends. *Journal of Polymer Science Part B: Polymer Physics, 38*(16), 2189–2211.
110. Visakh, P. M., Thomas, S., Chandra, A. K., & Mathew, A. P., (2013). *Advances in Elastomers, 1*, 1–23. Springer.
111. Visconte, L. L., Martins, A. F., Suarez, J. C., & Nunes, R. C., (2004). Different preparative modes for the incorporation of additives in NR/SBR blends. *Journal of Applied Polymer Science, 93*(2), 483–489.
112. Vo, L. T., & Giannelis, E. P., (2007). Compatibilizing poly(vinylidene fluoride)/nylon-6 blends with nanoclay. *Macromolecules, 40*(23), 8271–8276.
113. Wang, Q., Xie, J., Qiao, J., Zhang, L., Zhang, X., & Song, Z., (2012). Good dispersion of hydrophilic nanoscale silica in rubber matrix and the effects on rubber nanocomposites. *Science China Technological Sciences, 55*(3), 652–658.
114. Wu, W., & Chen, D., (2011). Silica-modified SBR/BR blends. *Journal of Applied Polymer Science, 120*(6), 3695–3700.
115. Yehia, A. A., Ismail, M. N., Hefny, Y. A., Abdel-Bary, E. M., & Mull, M. A., (2004). Mechano-chemical reclamation of waste rubber powder and its effect on the performance of NR and SBR vulcanizates. *Journal of Elastomers and Plastics, 36*(2), 109–123.
116. Zarei, M., Naderi, G., Bakhshandeh, G. R., & Shokoohi, S., (2013). Ternary elastomer nanocomposites based on NR/BR/SBR: Effect of nanoclay composition. *Journal of Applied Polymer Science, 127*(3), 2038–2045.

117. Zhang, H., (2009). *EPDM-Rubber in Blends with NR/BR-Elastomers for Ozone-Resistant Tire Sidewall Applications: New Approaches for Improved Mechanical Properties, 1*, 1–33.

118. Zhang, Y., Ge, S., Tang, B., Koga, T., Rafailovich, M. H., Sokolov, J. C., & Nguyen, D., (2001). Effect of carbon black and silica fillers in elastomer blends. *Macromolecules, 34*(20), 7056–7065.

CHAPTER 4

Mechanical and Physical Characteristics of Rubber Compounds

Contents

4.1. Introduction 160
4.2. Fatigue of Rubber 160
4.3. Crack Growth Examination and Fracture 163
4.4. Rubber Filler Interaction 166
4.5. Rubber Crystallization 168
4.6. Rubber Molecular Orientation 170
References 171

4.1. INTRODUCTION

Natural rubber (NR) is a material that is heavily researched. Aspects of NR that have usually been researched comprise: crack growth examination and fracture, the fatigue of rubber, rubber filler interactions, aging of rubber, rubber crystallization, rubber reversion, cord rubber composites, and molecular orientation of the rubber (Thomsen et al., 2001; Ermilov and Nurullaev, 2012). The evaluation of the examinations performed by some other researchers gives the foundation for this chapter. This literature evaluation will try to cover approximately all aspects of research conducted on NR along with other rubbery materials. The focus will be laid on the idea of energy, crack growth, temperature, and aging (Ferry et al., 1952; Meinecke and Taftaf, 1988).

4.2. FATIGUE OF RUBBER

An investigation into the wear out of rubber has concentrated on the procedure of fatigue failure along with on the growth of associations to envisage failure (Frogley et al., 2003; Tolliver and Clark, 2014). The work done by Rivlin et al.1 was quite influential in altering how the fatigue catastrophe of rubber was theorized. Rivlin et al.1 established the standard for tearing of the rubber hardness that was free of the shape of the piece of test. This form was centered on the failure theory of Griffith that defines the slow transmission of the crack as the transformation of elastic energy stowed in bulk to the surface free energy. Griffith suggested that at constant complete deformation, the variations in the energy stored elastically because of an upsurge in the length of crack were considered by the energy utilized in the formation of new surfaces. Though, Rivlin et al. (1948) suggested that under similar conditions, the variations in the energy stored elastically were balanced by variations in the internal energy. The energy was the work used irreversibly per cm upsurge in the length of the crack and per cm thickness of the piece of test. This standard of Rivlin et al. was thought the tearing energy principle, and it was exhibited that this distinctive energy could be discovered experimentally for a specific vulcanizate and utilized to forecast the force needed to tear the test pieces of the vulcanizate (Termonia, 1989; Böse and Röder, 2009).

Greensmith et al. (2005) has authenticated the tearing energy standard in the situation where a crack or cut is existent in the rubber sample strained in the simple extension (Figure 4.1). The criterion is given as (Reincke et al., 2009; Kuta et al., 2010):

$$T = -\left(\frac{1}{t}\right)\left(\frac{\partial U}{\partial c}\right)_l \qquad (1)$$

where; T is the characteristic tearing energy; t is the thickness of a test sample; l is the deformed length; c is the crack length; U is the elastically energy density from an imposed deformation.

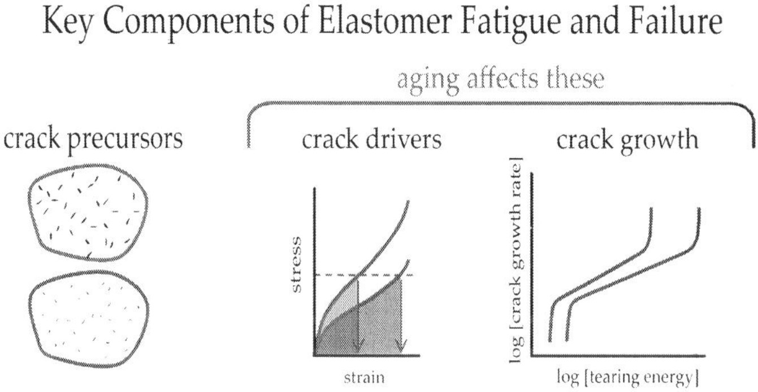

Figure 4.1. Graphic representation of crack propagation in rubbers.

Source: https://endurica.com/getting-a-quick-read-on-durability-with-the-intrinsic-strength-analyzer/.

Gent (1994) made practical utilization of the tearing energy standard to describe the association between fatigue and crack growth. Gent et al.3 discovered that the fatigue growth of crack of NR vulcanizates under recurrent loading for diverse test geometries was steady when inferred in expressions of the energy torn. The crack propagation per cycle was directly proportional to the square of the utmost tearing energy acquired during the cycle. Utilizing that expression, the fatigue lifespan of a sample containing the small crack was appraised from elasticity theory as the function of the preliminary size of the crack and the utmost applied strain. The strain dependency was also recognized for die-stamped samples that contained no deliberately induced flaws. To conclude that the tearing energy wasn't dependent on material, Lake (2000) protracted the utilization of tearing energy standard to the non-crystallizing rubbers like styrene-butadiene rubber (Hansen and Thomson, 2004; Özdemir et al., 2016).

The NR's mechanical fatigue life has been examined by Lake et al. (2000). In the research, they discovered that there prevails crucial tearing energy (T_0) for which the mechanico-oxidative growth of crack can take place. For the energies below this particular threshold, the growth of crack is accredited exclusively to chemical attack by the ozone. At the energies above this threshold, the growth of crack per cycle quickly increases and achieves the value of a few angstroms for each cycle. The rate of crack growth puts the growth of crack accomplished for each cycle on a similar scale as the chains of molecule. In the fatigue test, T_0 is the crucial value of strain below which the cycles number to failure upsurges quickly. This critical strain establishes the mechanical fatigue boundary for rubber (Kurian et al., 1987; Sircar et al., 1999).

Testing conditions like strain rate, applied strain level, oxidation effects, and testing temperature have been understood in combination with the fatigue crack growth (Lake, 1983). Young (1986) confirmed that with the increase in tearing energies, there can be a sharp transition from reasonable steady-state rates of crack growth to the disastrous crack growth area. This is comparable to the 3 areas of fatigue crack propagation, but the variation in tearing energy (ΔT) is used instead of the variation in strain energy release rate (ΔG). Increasing temperatures usually upsurges the slope of fatigue crack propagation, but this upsurge in slope can fluctuate greatly between diverse elastomers, e.g., SBR (styrene-butadiene rubber), BR (butadiene rubber), and NR. Both BR and NR exhibited slight strain rate dependence in the range 1–20 $sec.^{-1}$. Ozonation and oxidation greatly exaggerated the fatigue crack growth, where the addition of 0.5 ppm (parts per million) ozone to the air atmosphere augmented the crack growth rate by 40–80% compared to samples in the air atmosphere (Gehman et al., 1947; Kern and Futamura, 1988).

Recently, Bathias et al. (1987) offered his work on fatigue instigation and propagation in NR and SR with respect to the chemical microstructure of a rubber, environment, loading, thickness, and intricate loadings. Bathias et al. demonstrated that fatigue damage is dependent on 3 basic factors:

- Chemical composition;
- Environment; and
- Mechanical stresses.

In the experiments, axisymmetric hour-glass samples were loaded in compression and tension. Fatigue behavior was enhanced for tests with the mean stress in some tension, $R \geq 0$, whereas minimum stress in the compression, R < 0, significantly increased the damage of fatigue. In the research, Bathias et al. established that damage because of chemical effects are quite greater at higher temperatures because of the acceleration of oxidation reaction. Additionally, Bathias et al. exhibited that the fatigue growth of crack of the rubbers can generally be modeled by application of the fracture mechanics. They also discovered that cavitation happened in the interior of a specimen because of high hydrostatic pressures. The cavitation mechanism happened under cyclic tension, static tension, and the static tension torsion loading (ASTM D573, 2004).

The exploration into the lassitude of elastomers has exhibited that the fatigue growth of crack of the rubbers can usually be modeled with the help of fracture mechanics. The method of tearing energy has generally been broadly applied and acknowledged to define rubber fatigue crack propagation. The crucial tearing energy has been discovered to prevail below which all of the growth of crack is accredited to oxidation. For the energies above this threshold, crack growth upsurges, ultimately obtaining the rate on a scale of just 1 polymer monomer unit for each cycle. The fatigue lifespan instigation and propagation in synthetic and NRs are usually reliant on the environment, chemical composition, and mechanical stresses given to the sample (Edwards and Storey, 1955; Standard, 2012).

4.3. CRACK GROWTH EXAMINATION AND FRACTURE

The utilization of tearing energy standards to the growth of crack and fracture of NR has been executed by several researchers. Lake et al. (1991) has displayed that the crack growth rate is free of specimen shape and kind of deformation when inferred in terms of elastic energy. The work was meant to determine the procedure of groove cracking in the treads of the tire. Lake (2000) applied the technique which empowered the elastic energy in order to be assessed for the groove of the tire from the measurements of the crack opening region upon the stress applied. The technique also allowed Lake et al. (1991) to conclude the strain inside the sample. Lindley (1972) examined three techniques of defining the tearing energy: from the crack surface displacements, from variations in total energy, and by comparing with already identified values for the similar rates of crack growth. It was discovered that all of the three techniques, when applied to the plane stress samples, were acceptable in concluding the energy for growth of the crack in the model rubber components. Mainly, the crack surface displacement technique, analogous to Lake et al. (2012), was discovered to be of utmost significance (Crozier et al., 1974; Ismail et al., 2011).

The velocities of a crack in NR have normally been estimated by Gent et al. (1994). They examined biaxially strained sheets of carbon black (CB) filled and unfilled NR. The velocity of crack was discovered to upsurge with the strain, ε_x, with the and cleavage strain, ε_y, parallel to the direction of tearing. The velocities over 100 m/sec were acquired, 1/3rd the speed of the sound. It was discovered that below the threshold value of cleavage strain-no growth of crack occurred. This restraining behavior is comparable to critical tearing energy as explained by Lake et al. (1991). Gent et al. (1994) did observe a dissimilarity in the behavior amongst the unfilled and filled materials. For the material filled with CB, cracks grew at a very low velocity for the series of strains normally above the threshold strain, and then the velocity increased rapidly. Though, for an unfilled material, the cracks grew quickly once the critical strain was exceeded. The velocities of high-speed cracks were discovered to be free of the accessible elastic energy for the fracture and depend on the shape of the biaxial strain. Therefore, filler has an impact upon the rate of crack propagation in NR; though these rates aren't solely the function of elastic energy, the applied deformations also.

Rivlin (1948) observed the influence of stress relaxation on-resistance of the elastomer to tearing. Rivlin (1948) used the two-network model for the elastomers in which the stress relaxation has occurred already. Rivlin (1948) observed the conformation of an elastomer when distorted as being in two diverse states. The two different states happen because, under the large stress applied, recombination, and rupture of crosslinks can occur. Such large stress applied can normally be in an area near the tip of the crack. Therefore, after relaxation, these two configurational states are normally in the unstressed state. One of the configurational states is centered on the original conformation of elastomers, whereas the other state is centered on the state of deformation applied. When the distorting forces are eliminated, the elastomeric network will embrace the configurational state which is dissimilar from the preliminary state and is amongst the unstressed state of two-component networks (Shaffer et al., 2008; Atashi and Shiva, 2010).

The method of Rivlin et al. (1948) comprised the influence of the relaxation on an elastomer, in the background of tearing energy, made practical to the supposition of the kinetic theory of the rubber elasticity. The tearing energy was measured utilizing the approach of Lake et al. (1991), as the energy needed to break the chain segment. Lake et al. (1991) supposed the isotropic elastomer in which the chain segment should be extended fully to break. Whereas just one chemical bond is required to break to trigger failure, all chemical bonds should be stretched to almost the point of breaking. Rivlin et al. (2016) discovered that whether the chain segments were supposed to be non-Gaussian or Gaussian or not, a considerable extent of anisotropy in tearing energy is anticipated depending on the ratio of two component networks. The characteristic tearing energy was discovered to be better parallel to the orientation of extension instead of perpendicular when the number of links for each chain segment was nearly equal in the two networks. Though, if the number of links for each chain segment was more in the network centered on the state of deformation applied, the characteristic tearing energy was generally less in the orientation parallel to an extension. These two effects were normally greater for a non-Gaussian model (Alexander and Thachil, 2006).

The idea of tearing energy has usually been extended utilizing the J-integral (Lutfi et al., 2020). The tearing energy and the J-integral methods describe the fracture properties by supposing that the crucial sum of energy is needed to advance the crack by unit region. Kuai et al. (2009) discovered that the J-integral method showed less scatter as compared to the characteristic tearing energy but needed the larger length of crack versus length of sample ratio to acquire accurate values. Moreover, the J-integral method separated the crucial energy in an area of the tip of the crack and the stowed elastic energy in the majority. Kuai et al. (2009) suggested that the rate of crucial crack tip energy release for instigation and fracture be deliberated material properties, which signify the materials confrontation to fracture along with being a suitable parameter for the evaluation of material (Hwang et al., 2002).

Instead of utilizing the single edge crack samples, which are restricted in the array of achievable energies, new methods to measuring the growth of the crack in NR are being made. Sumner et al. (1995) has carried out research utilizing the tear analyzer. The machine automatically calculates the rate of crack propagation in rubber and also mechanical parameters during the dynamic testing. The crack analysis of the unfilled NR utilizing infrared microspectroscopy has generally been examined by Neumeister and Koenig (1996). This method demonstrated that in the strained crack the material neighboring the tip of the crack possesses different features as compared to bulk. In the test of razor cut the tip of the crack region of the unstressed material was identical spectroscopically to the material in the majority (Medalia, 1978).

Gent (1994) have examined the rates of energy release acquired from the crack in the tilted block. They also discovered that the thin long block is probable to suffer internal cavitation before achieving adequate fracture energy in order to propagate the crack inwards. The cavitation is an outcome of hydrostatic tension produced inside the rubber because of the load applied Gent (1994) discovered that long thin blocks are more appropriate for the growth of crack measurements since higher values of the rate of energy release can be accomplished without triggering cavitation. The inspiration for this exploration was the propagation of the crack sample observed in this chapter.

4.4. RUBBER FILLER INTERACTION

The collaboration between the filler and the rubber has been studied in order to establish the impacts on the failure of compounds. Bhaumik et al. (1988) have investigated the high-temperature collaboration between filler and rubber by utilizing the factor of strain amplification. When the CB is normally added for strengthening, the extent of freedom of rubber chains is reduced because of the collaboration and adsorption of the non-deformable CB. Upon the load applied, the rubber should bear the entire strain; though, the local strain insider the rubber phase is far greater as compared to the global strain acquired by the system (Freakley and Payne, 1978; Edwards, 1983).

The difference between the global and the local strains is known as the factor of strain amplification. Bhaumik et al. (1988) established the factor of strain amplification at different temperatures and strain rates with N-660, N-550, N-375, N-220 CBs, and the china clay as a filler at 50 pph (parts per 100) parts polymer.

For all of the specimens, Bhaumik et al. (1988) discovered that the factor of strain amplification accomplished the restraining value at 100°C. The value was far greater for fillers having a higher surface area, therefore producing the greater interaction with a rubber (Figure 4.2) (Gent, 1994).

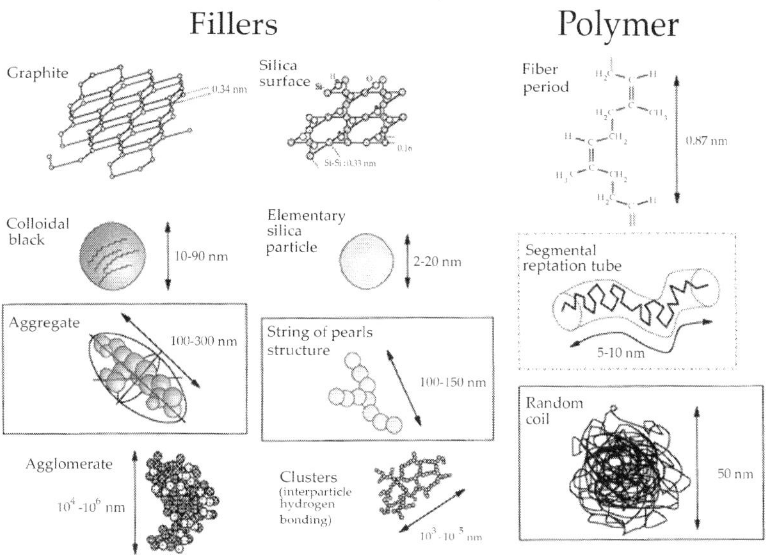

Figure 4.2. Rubber-filler collaborations.

Source: https://www.sciencedirect.com/science/article/pii/S0079670001000405.

Chung et al. (1982) examined the impacts of the CB on critical properties of the elastomers. They observed the To (critical tearing energy) along with the crucial J-integral. They discovered that for NR compounds, crystallization might be observed close to the tip of the crack. Crystallization hindered the propagation of crack over the thickness of the sample. The CB loading level reasonably affected the critical tearing energy of the NR compounds, with values varying from 4.1 kN/m–6.9 kN/m. The modulus of the compound was discovered to upsurge with increasing CB levels.

Wang et al. (2002) have carried out a comprehensive investigation of the impacts between filler-filler and rubber-filler on the vibrant properties of vulcanizates. Analogous to Chung et al. (1982); Wang et al. (2002) discovered that the true modulus of compound upsurged with increasing CB levels. The upsurge was constant for the loss modulus along with the storage modulus. Wang26 attributed the stiffening because of the filler by adsorption of the polymer chains of a molecule on the surface of the filler. This adsorption decreases the mobility of polymer segments and outcomes in the shell of rubber on the surface of the filler. The decreased mobility and the shell of rubber increase the viscosity of the polymer. This upsurge in viscosity formed the broadening of the spectrum of the relaxation times and the modulus increases accordingly.

The collaboration between the filler and the rubber disturbs the quasi-static properties of compounds along with the fatigue life. Insertion of the CB into NR compounds acts to upsurge the modulus of compound along with hinder propagation of the crack. This decrease in the propagation of crack occurs due to localized crystallization happening in the compound because of regions of highly controlled polymer. These regions exist near the tip of a crack along with the regions of high limitation because of the filler particles (Boretos and Pierce, 1968; Engel et al., 2006).

4.5. RUBBER CRYSTALLIZATION

This occurs because of the reduction in localized entropy upon the forced deformation (Goritz & Kiss, 1985). This event is described by Gibbs Free Energy. The thermodynamic formula of the Gibbs Free Energy is given:

$$\Delta G = \Delta H - T\Delta S \quad (2)$$

where; ΔG is the variation in Gibbs free energy; ΔH is the variation in enthalpy; T is the absolute temperature; ΔS is the variation in entropy of the system.

At equilibrium, in a liquid state, where generally $T > T_g$, the variation in free energy is 0 and the variation in enthalpy is equal to enthalpy of the fusion, ΔH_f. In the equation above, T then becomes the system's local melting temperature, T_m. The entropy by an equation of the Boltzmann is given as:

$$S = k \ln \Omega \quad (3)$$

where: k: Boltzmann's constant; Ω: Number of the configurational states of polymer.

Under the forced malformation, the number of the configurational states approaches 1. This configuration variation happens for polymeric chains lining up themselves in the orientation of the load applied. Therefore, ΔS approaches 0. To maintain the continuity of Eqn. (2), the T_m must increase. Once the T_m upsurges below the surrounding testing temperature crystallization will take place (Figure 4.3) (Mix and Giacomin, 2011; Ajam et al., 2016).

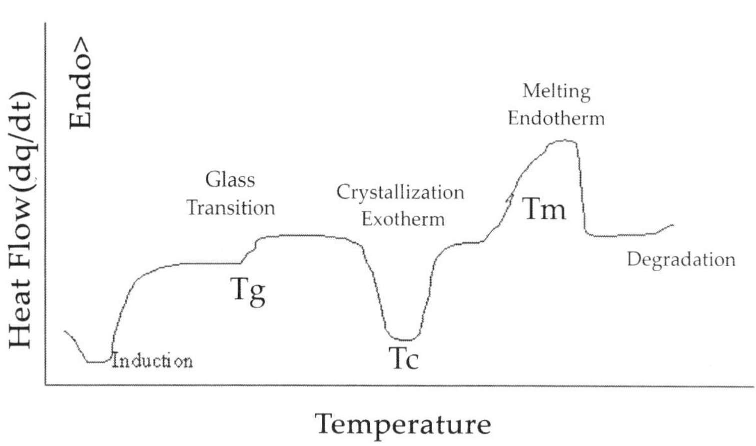

Figure 4.3. Differential skimming calorimetry graph of the general rubber compound.

Source: https://www.pslc.ws/macrog/mpm/analysis/dsc.htm.

Allegra and Bruzzone (1983) have discovered that the rubber crystallization can usually be modeled as the third-order transition. This transition is different as compared to Flory's (1944) second-order theory. Goritz and Kiss (1985) discourses an additional procedure of strain-induced crystallization. If the utmost in the crystallinity's degree isn't reached under the load applied, then the residue of crystallizable chains crystallize on decreasing the temperature. Goritz and Kiss (1986) examined the two strain-induced crystallization event by carrying out differential skimming calorimetry scans on malformed samples. For the cis-1,4 polybutadiene, sample extended to 400% strain the two crystalline melting areas appeared distinctly. The full width half maximum (FWHM) of the temperature tempted crystalline region was around 267 K whereas the FWHM of the strain tempted crystalline region was nearly 310 K. The FWHM of the temperature tempted crystallization between the unstrained sample and a strained sample varied by 5 K. Goritz and Kiss (1986) described this difference as the stress tempted entropic effect.

The crystallization in NR is the stress tempted entropic effect. The decrease in entropy can happen in the areas of high-stress concentrations like the crack tip. The collaboration between filler and rubber can disturb the NR's mechanical properties, but what impacts are there to a compound as the function of time (Lu et al., 2004; Schaefer, 2010).

4.6. RUBBER MOLECULAR ORIENTATION

The rubber's molecular orientation is normally examined in terms of the load applied and the subsequent strain tempted crystallization. Techniques like wide-angle x-ray scattering. Fourier transform IR spectroscopy., stationary fluorescence polarization and deuterium magnetic resonance (Long, 1985; Kawahara et al., 2019). have been applied in order to conclude the direction parameters of a network. Though, when comprehending polymers, one should keep in the notice the history of a sample before testing. Whereas the rubber elasticity theory is centered on an irregular, 3-D network, it doesn't consider any effects of history. At room temperature, there isn't any physical aging of the NR vulcanizate, as $T_g \approx -75°C$, but there exists a mechanical history of the specimen (Donnet and Voet, 1976; Foster, 1987).

REFERENCES

1. Ajam, A. M., Al-Nesrawy, S. H., & Al-Maamori, M., (2016). Effect of reclaim rubber loading on the mechanical properties of SBR composites. *International Journal Chemist. Sci., 14*(4), 2439–2449.
2. Alexander, M., & Thachil, E. T., (2006). A comparative study of cardanol and aromatic oil as plasticizers for carbon-black-filled natural rubber. *Journal of Applied Polymer Science, 102*(5), 4835–4841.
3. Allegra, G., & Bruzzone, M., (1983). Effect of entropy of melting on strain-induced crystallization. *Macromolecules, 16*(7), 1167–1170.
4. ASTM D573, (2004). *Standard Test Method for Rubber-Deterioration in an Air Oven, 1*, 1–10.
5. Atashi, H., & Shiva, M., (2010). Formulation for passenger tread tire compound based on styrene-butadiene rubber/cis-butadiene rubber/natural rubber blend and semi-dispersible silica/carbon black filler system. *Asian Journal of Chemistry, 22*(10), 7519.
6. Bathias, C., (1999). There is no infinite fatigue life in metallic materials. *Fatigue and Fracture of Engineering Materials and Structures (Print), 22*(7), 559–565.
7. Bhaumik, T. K., Neogi, C., Gupta, B. R., & Bhowmick, A. K., (1988). Abrasion of high temperature conveyor belt compounds based on ethylene propylene diene and bromobutyl rubber blends. *Wear, 128*(2), 167–178.
8. Boretos, J. W., & Pierce, W. S., (1968). Segmented polyurethane: A polyether polymer. An initial evolution for biomedical applications. *Journal of Biomedical Materials Research, 2*(1), 121–130.
9. Böse, H., & Röder, R., (2009). Magnetorheological elastomers with high variability of their mechanical properties. In: *Journal of Physics: Conference Series* (Vol. 149, No. 1, p. 012090).
10. Chung, K. T., Sabo, A., & Pica, A. P., (1982). Electrical permittivity and conductivity of carbon black-polyvinyl chloride composites. *Journal of Applied Physics, 53*(10), 6867–6879.
11. Crozier, W. F., Stoker, J. R., Martin, V. C., & Nordlin, E. F., (1974). Laboratory evaluation of full-size, elastomeric, bridge bearing pads. *Concrete, 55*, 1–33.
12. Donnet, J. B., & Voet, A., (1976). *Carbon Black: Physics, Chemistry, and Elastomer Reinforcement* (Vol. 1, pp. 1–22). M. Dekker.

13. Edwards, D. C., & Storey, E. B., (1955). Ozone resistance of butyl vulcanizates. *Rubber Chemistry and Technology, 28*(4), 1096–1118.
14. Edwards, R. C., (1983). *Automotive Elastomers and Design, 1*, 1–22.
15. Engel, J. M., Chen, J., Liu, C., & Bullen, D., (2006). Polyurethane rubber all-polymer artificial hair cell sensor. *Journal of Microelectromechanical Systems, 15*(4), 729–736.
16. Ermilov, A. S., & Nurullaev, E. M., (2012). Mechanical properties of elastomers filled with solid particles. *Mechanics of Composite Materials, 48*(3), 243–252.
17. Ferry, J. D., Fitzgerald, E. R., Grand, L. D., & Williams, M. L., (1952). Temperature dependence of dynamic mechanical properties of elastomers, relaxation distributions. *Industrial and Engineering Chemistry, 44*(4), 703–706.
18. Flory, P. J., (1944). Network structure and the elastic properties of vulcanized rubber. *Chemical Reviews, 35*(1), 51–75.
19. Foster, V. R., (1987). Polymers in caulking and sealant materials. *Journal of Chemical Education, 64*(10), 861.
20. Freakley, P. K., & Payne, A. R., (1978). *Theory and Practice of Engineering with Rubber, 1*, 1–22. Applied Science Publishers.
21. Frogley, M. D., Ravich, D., & Wagner, H. D., (2003). Mechanical properties of carbon nanoparticle-reinforced elastomers. *Composites Science and Technology, 63*(11), 1647–1654.
22. Gehman, S. D., Woodford, D. E., & Wilkinson, Jr. C. S., (1947). Low temperature characteristics of elastomers. *Industrial and Engineering Chemistry, 39*(9), 1108–1115.
23. Gent, A. N., (1994). *How to Design Rubber Components* (Vol. 1, pp. 1–22). Pennsylvania 16514–0039, USA.
24. Goritz, D., & Kiss, M., (1986). On the origin of strain-induced crystallization. *Rubber Chemistry and Technology, 59*(1), 40–45.
25. Greensmith, J., Aickelin, U., & Cayzer, S., (2005). Introducing dendritic cells as a novel immune-inspired algorithm for anomaly detection. In: *International Conference on Artificial Immune Systems* (Vol. 1, pp. 153–167). Springer, Berlin, Heidelberg.
26. Hansen, P., & Thomson, B., (2004). The development and exploitation of accelerated durability tests: The new ASTM D429 method G immersion test and potential future developments. In: *Polymer Bonding*

2004 (Bonding of Rubber and Plastics to Various Substrates) (Vol. 1, pp. 51–65).

27. Hwang, J. S., Wu, J. D., Pan, T. C., & Yang, G., (2002). A mathematical hysteretic model for elastomeric isolation bearings. *Earthquake Engineering and Structural Dynamics, 31*(4), 771–789.
28. Ismail, E. A., Motawie, A. M., & Sadek, E. M., (2011). Synthesis and characterization of polyurethane coatings based on soybean oil-polyester polyols. *Egyptian Journal of Petroleum, 20*(2), 1–8.
29. Kawahara, S., Yamamoto, Y., & Saito, T., (2019). Rubber-state NMR Spectroscopy. *NMR Methods for Characterization of Synthetic and Natural Polymers, 20*, 156.
30. Kern, W. J., & Futamura, S., (1988). Effect of tread polymer structure on tire performance. *Polymer, 29*(10), 1801–1806.
31. Kuai, H., Lee, H. J., Zi, G., & Mun, S., (2009). Application of generalized J-integral to crack propagation modeling of asphalt concrete under repeated loading. *Transportation Research Record, 2127*(1), 72–81.
32. Kurian, J., Nando, G. B., & De, S. K., (1987). Scanning electron microscopic studies on the failure of rubber-to-metal bonded composites. *The Journal of Adhesion, 20*(4), 293–316.
33. Kuta, A., Hrdlicka, Z., Voldanova, J., Brejcha, J., & Pokorny, J., (2010). Dynamic mechanical properties of rubbers with standard oils and oils with low content of polycyclic aromatic hydrocarbons. *KGK. Kautschuk, Gummi, Kunststoffe, 63*(4), 120–122.
34. Lake, C., (2000). *U.S. Patent No. 6,050,903* (Vol. 1, pp. 1–22). Washington, DC: U.S. Patent and Trademark Office
35. Lake, G. J., Samsuri, A., Teo, S. C., & Vaja, J., (1991). Time-dependent fracture in vulcanized elastomers. *Polymer, 32*(16), 2963–2975.
36. Lindley, P. B., (1972). Energy for crack growth in model rubber components. *Journal of Strain Analysis, 7*(2), 132–140.
37. Long, H., (1985). *Basic Compounding and Processing of Rubber* (Vol. 1, pp. 1–33). Rubber Division, American Chemical Society Incorporated.
38. Lu, H., Nguyen, B., & Powers, J. M., (2004). Mechanical properties of 3 hydrophilic addition silicone and polyether elastomeric impression materials. *The Journal of Prosthetic Dentistry, 92*(2), 151–154.

39. Lutfi, S. N. N., Abd, R. N. A., Ali, S., & Gholizadeh, H., (2020). Compression and tension behavior of the prosthetic foam materials polyurethane, EVA, pelite™ and a combination of polyurethane and EVA: A preliminary study. *Biomedical Engineering/Biomedizinische Technik, 1, 1–22.*

40. Medalia, A. I., (1978). Effect of carbon black on dynamic properties of rubber vulcanizates. *Rubber Chemistry and Technology, 51*(3), 437–523.

41. Meinecke, E. A., & Taftaf, M. I., (1988). Effect of carbon black on the mechanical properties of elastomers. *Rubber Chemistry and Technology, 61*(3), 534–547.

42. Mix, A. W., & Giacomin, A. J., (2011). Standardized polymer durometry. *Journal of Testing and Evaluation, 39*(4), 696–705.

43. Özdemir, T., Akbay, I. K., Uzun, H., & Reyhancan, I. A., (2016). Neutron shielding of EPDM rubber with boric acid: Mechanical, thermal properties and neutron absorption tests. *Progress in Nuclear Energy, 89*, 102–109.

44. Reincke, K., Grellmann, W., & Friedel, J., (2009). Influence of process oils on the mechanical properties of elastomers. *Kautsch Gummi Kunstst, 62*(10), 506–514.

45. Rivlin, R. S., (1948). Large elastic deformations of isotropic materials IV. Further developments of the general theory. *Philosophical Transactions of the Royal Society of London. Series A, Mathematical and Physical Sciences, 241*(835), 379–397.

46. Schaefer, R. J., (2010). Mechanical properties of rubber. In: Piersol, A., & Paez, T., (eds.), *Harris' Shock and Vibration Handbook* (6th edn., Vol. 1, pp. 30–33). McGraw-Hill Companies Inc.

47. Shaffer, G., Sanchez, M. R., & Rosenberg, B., (2008). The trials of winning at the WTO: What lies behind Brazil's success. Cornell Int'l LJ, 41, 383.

48. Sircar, A. K., Galaska, M. L., Rodrigues, S., & Chartoff, R. P., (1999). Glass transition of elastomers using thermal analysis techniques. *Rubber Chemistry and Technology, 72*(3), 513–552.

49. Standard, A. S. T. M., (2012). Standard test method for rubber property-effect of liquids. *ASTM D471, 1, 1–22.*

50. Termonia, Y., (1989). Molecular model for the mechanical properties of elastomers. 1. Network formation and role of entanglements. *Macromolecules, 22*(9), 3633–3638.
51. Thomsen, D. L., Keller, P., Naciri, J., Pink, R., Jeon, H., Shenoy, D., & Ratna, B. R., (2001). Liquid crystal elastomers with mechanical properties of a muscle. *Macromolecules, 34*(17), 5868–5875.
52. Tolliver, C., & Clark, D., (2014). Identifying important variables in compression set testing. *Sealing Technology, 2014*(5), 7–11.

CHAPTER 5

Conductive Elastomers and Their Electronic Applications

Contents

5.1. Introduction....178

5.2. Stretchable Electronics....180

5.3. Stretchable Sensors....188

5.4. Stretchable Energy Harvesters....194

5.5. Summary....198

References....199

5.1. INTRODUCTION

There had been an extensive variety of exertions to develop conductive elastomers that fulfill both electrical conductivity and mechanical stretchability, as a response to increasing demands on wearable devices. This chapter analyzes the significant development in conductive elastomers completed in 3 application fields of stretchable technology: stretchable sensors, stretchable energy harvesters, and stretchable electronics. Different combinations of non-stretchable conductive materials and insulating elastomers had been studied to understand optimal conductive elastomers (Kim and Rogers, 2008; Ahn and Je, 2012). It is noted that similar structures and similar material combinations had frequently been employed in diverse fields of application. In terms of cyclic operation, stretchability, and overall performance, fields like as stretchable pressure/strain sensors and stretchable conductors had attained great advancement, while other fields like stretchable thermoelectric energy harvesting and stretchable memories are in their beginning. It is worth stating that there are still problems to overwhelmed for the further development of stretchable technology in the relevant fields, which comprise the device structure and simplification of material combination, reliability, and securement of reproducibility, and the formation of easy fabrication methods. By this review article, both the obstacles and progress related with the relevant stretchable technologies would be understood more obviously (Someya et al., 2005; Rogers et al., 2010).

The demand for stretchable devices had been ever-rising as novel technology fields like intelligent robotics, body conformable devices, stretchable electronics, and wearable devices, had emerged (Kim and Rogers, 2008). For example, keen sensory skins are needed to execute advanced robots that could interact well with humans and appropriately respond to the environment without exterior control (Chou et al., 2015). Keeping pace with this increasing need for novel technology, a movement of searching for novel materials that could afford good mechanical elasticity and also high electrical conductivity had surged (Faez et al., 2002). Though a diversity of conducting polymers, like polypyrrole (PPY), polyaniline (PANI), polyacetylene (PA), and poly(3,4-ethylene dioxythiophene (PEDOT) had been developed for various applications, their extensive usage is limited through their poor mechanical properties. For instance, a (PEDOT: PSS) poly(3,4-ethylene dioxythiophene): poly(styrene sulfonic acid) film, which is extensively employed in organics founded optoelectronic devices and plastic electronics, displays high electrical conductivity up to 1000 S/cm;

however, its breaking strain is under 10% (Lang et al., 2009). This level of forbearance to strain is not suitable for the above-stated applications.

On the contrary, elastomers like SBR (styrene-butadiene rubber), polyurethane (PU), poly(dimethylsiloxane (PDMS), natural rubber (NR), and ethylene-propylene-diene monomer (EPDM) are characterized by poor conductivity, high, reversible deformation (>200%) (Kim et al., 2011). Traditionally, they had been utilized mainly for structural, industrial, and household products, and numerous fillers had been integrated into them to fortify mechanical properties like Young's modulus and tensile strength (Liu et al., 2015). Excitingly, certain conductive fillers characterized by a family of carbon matters, like graphites, carbon blacks (CBs), and carbon nanotubes (CNTs) had been introduced to transmute the resin from an insulator to a conductor. Though this approach needs a substantial amount of filler to be added, occurring an extreme loss of the material's elasticity; however, the degradation of elasticity could be reduced for CNT-elastomer mixtures, where conductive CNT categorization networks could be formed through the addition of just a little amount of CNTs (Benight et al., 2013; Chortos and Bao, 2014).

For the sake of the instantaneous gratification of good elasticity and high electrical conductivity, certain groups had attempted to alter the molecular structures of elastomers. Doping SBR with antimony pentachloride ($SbCl_5$) and iodine (I_2), and producing graft copolymers composed of PANI and PU are typical instances (Abbati et al., 2003). Though, these methods need elaborate modification of experimental conditions and are still to endorse their long-term reliability. A more practical and easy method of tackling the goal is to create a blend comprising of an elastomer and a conducting polymer (Hansen et al., 2007). In this structure, the conducting polymer shows a role in enhancing the conductivity of the mixture; however, the elastomer condenses the material stretchable. When 11.5% of PANI by volume portion was added to PU, the material's conductivity enhanced by six orders of magnitude, however, keeping a high stretchability of 200%. For a PEDOT-PU merger, a high conductivity of 100 S/cm was revealed, even under an elongation of >100%. Usually, the electrical conductivity of a blend rises at the cost of mechanical properties as the portion of the conductive part enhances. Therefore, elaborate material design is essential to accomplish detailed specifications needed for precise applications (Wang et al., 2001; Bhadra et al., 2009).

In this review, various exertions to realize conductive elastomers and the latest progress are debated. Owing to the vastness of earlier works, some representative outcomes in the fields of sensors, electronics, and energy harvesting would be introduced. Notably, similar material combinations and similar strategies had sometimes been accepted in diverse fields of application. Although this review, not only accomplishments made till yet, however tasks to be solved would also be deliberated (Tamai, 1982; Chou et al., 2015).

5.2. STRETCHABLE ELECTRONICS

The scope of stretchable electronics is very extensive, comprising memories, stretchable logic gates, and stretchable display units. Although the relevant devices had diverse structures, they are usually organized with basic elements, like transistors, interconnects, dielectrics, and light-emitting diodes (LEDs). One of the significant strategies for executing stretchable electronics is a mixture of active, rigid components, and stretchable interconnects, which accommodate exterior strains (Lacour et al., 2004; Sun et al., 2006). Although to realize actual stretchable electronics, entire components are required to be stretchable. From the part of materials, this needs semiconductors, conductors, and dielectrics, all of them are stretchable. In this portion, stretchable conductors and certain representative stretchable devices are concisely reviewed (Faez et al., 2002; Saleem et al., 2010).

5.2.1. Stretchable Conductors

Two kinds of metal structures (*for example.*, serpentine structures and wavy structures) had been extensively explored as stretchable conductors. Wavy metal structures are made from metal films placed on the pre-patterned or strained elastomeric substrate (Kim et al., 2008), and their alterable stretchability is comparatively low (<20%). On the other side, serpentine metal structures are 2D or 3D ranges of a reiterating horseshoe-shaped unit. Hsu *et al.* confirmed the ultimate elongation of 250% and the reliability of 40,000 cycles at a 30% elongation, utilizing polyimide enhanced Cu serpentine structures (Hsu et al., 2011; Hilbich et al., 2015). Though these techniques proved their efficiency, they usually need a multi-step procedure comprising a costly lithography step (Das and Prusty, 2012; Balint et al., 2014).

Another method is to employ elastomeric compounds. The elastomeric compounds might be elastomer conducting polymer mixtures, elastomeric compounds incorporating metal nanostructures or carbon nanostructures, and polymer blends integrating nanostructures of the conductive materials (Faez et al., 1999; Stoyanov et al., 2013). Nearly two decades before, Fu et al. (1998) made elastomeric PU—PPY compound foams through *in situ* polymerization of pyrrole in pre-formed PU foams, and conductivity of 10^{-5} S/cm and elongation at the halt of 160% was attained from a 6 wt.% PPY comprising composite (Lee and Khang, 2012; Chen et al., 2013). The less conductivity of the compound, which was one of the major problems with techniques utilizing PU foams, had been improved through introducing conductive nanostructures. For instance, Ge *et al.* made-up Ag nanowire PU sponge PDMS stretchable conductors through a simple solution-dipping technique, and definite high conductivity of 19.2 S/cm, and a resistance change $(\Delta R/R_0)$ of 160% at a 100% strain (Ge et al., 2013). Rather than PU foams, Park et al. (2012) created 3D PDMS structures utilizing proximity field nanopatterning, and formed transparent, stretchable conductors through infiltrating liquid metal, EGaIn (eutectic gallium-indium), into the 3D PDMS (Figure 5.1(a)–(c)). The 3D PDMS-EGaIn stretchable conductors presented tremendously high conductivity (24,100 S/cm), even below strains of >200%, and good stretching cycle achievement (Figure 5.1(d) and (e)) (El-Tantawy et al., 2009; Khan et al., 2010).

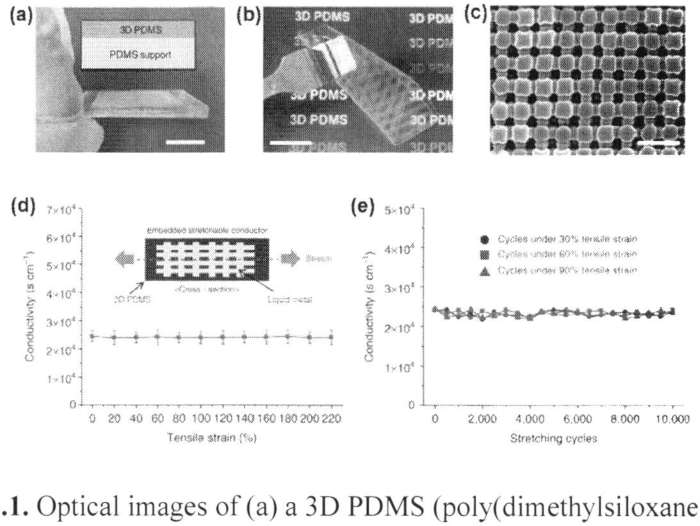

Figure 5.1. Optical images of (a) a 3D PDMS (poly(dimethylsiloxane) film on PDMS support; and (b) a folded 3D PDMS film. Scale bar, 1 cm; (c) top view SEM image of net made 3D PDMS film. Scale bar, 1 μm; (d) conductivity of 3D PDMS-EGaIn (eutectic gallium-indium) stretchable conductor below strains of

up to 220%; (e) conductivity variation relying on the number of stretching-releasing cycles under diverse strains.

Source: https://www.scholars.northwestern.edu/en/publications/three-dimensional-nanonetworks-for-giant-stretchability-in-dielec.

Hansen et al. (2007) made PU-PEDOT combinations from liquid mixtures of EDOT and fluctuating amounts of PU liquefied in tetrahydrofuran (THF) without involving porous elastomers (Rogers and Huang, 2009). They stated a good conductivity of 10–50 S/cm at a 200% strain for the mixtures. As a similar approach, Noh (2014) made PDMS-PEDOT: PSS combinations through introducing a miscibility increasing copolymer, poly(dimethylsiloxane-*b*-ethylene oxide (PDMS-*b*-PEO), and revealed a conductivity up to 2 S/cm and a fracture strain of 75% (Noh, 2014). Single-walled CNTs (SWCNTs) and multi-walled CNTs (MWCNTs) had been very extensively employed to transmute insulating elastomers to conductors. Kim et al. (2014) made SWCNT-PDMS composites through backfilling SWCNT aerogels and noticed conductivities of 70–108 S/m and a little resistance alteration of 14% at a tensile strain of 100%. Shin et al. (2010) used the same approach, in which allied MWCNTs were first made through catalyst-assisted chemical vapor deposition (CVD) and later infiltrated through PU solution. The resulting MWCNT-PU composites presented a conductivity of 50–100 S/m and alterable resistance variation for strains up to 40%. Graphene, additional material in the carbon family, had been progressively applied for stretchable conductors by a smart combination with suitable elastomers (Kuilla et al., 2010; Xu and Zhu, 2012). According to Lee et al. (2012) for example, a composite occupied of PU and functionalized graphene sheets could approach an elongation at a break of 374%, however retaining a conductivity of 1.2×10^{-5} S/cm.

Silver (Ag) nanostructures had also been intensively examined as a conductivity boosting factor for stretchable conductors. Zhu and Xu (2012) reported that their AgNWs (Ag nanowires) inserted PDMS composite attained a high conductivity of 5285 S/cm in a stretchable strain range of 0% to 50%. Lee et al. (2012) made networks of very lengthy AgNWs on Ecoflex utilizing a vacuum filtration and transfer technique, and they demonstrated high transparency of 90% to 96%, the sheet resistance of 9–70 Ω/sq, and good stretchability of >460%. Araki et al. (2011) formed Ag flakes PU composites through emulsion mixing and attained high stretchability up to 600% and low resistivity of 2.8×10^{-4} Ω¨ cm. In another interesting method, PU gold nanoparticle (AuNP) composites were prepared through vacuum-assisted flocculation or layer-by-layer assembly, and they presented an extreme conductivity of 11,000 S/cm and stretchability of 486% (Kim et al., 2013). Furthermore, AuNPs in these composites could be restructured under stress, permitting electronic regulation over mechanical properties (Sadhu and Bhowmick, 2004; Lang et al., 2009).

5.2.2. Stretchable Field-Effect Transistors and Memories

The common techniques of fabricating stretchable FETs (field-effect transistors) comprise the combined usage of stretchable conductors and rigid gate stack, implementation of wavy structures of the inorganic materials, and realization of composite FETs or organically made of entirely stretchable components. Shin et al. (2011) fabricated FET series of suspended SnO_2 NWs with wavy intersects and revealed high stretchability up to 40% and current on/off ratios of 10^6. Sekitani et al. (2008) fabricated a huge area stretchable active matrix comprising 19 × 37 organic transistors, integrating an SWCNT-founded elastomeric conductor. The device could be extended up to 70% both biaxially and uniaxially without mechanical fracture. Moreover, Kim *et al.* formed stretchable CMOS (complementary metal-oxide-semiconductor) inverters and 3 step ring oscillators on PDMS. Those devices implementing wavy structures of single-crystalline Si (silicon) nanoribbons presented stable oscillation frequency of ~3 MHz and high gains of 100, even under a 5% strain (Ghosh and Chakrabarti, 2000; Kojio et al., 2010).

Lately, Jeong's group developed stretchable transistors prepared completely of stretchable components (Shin et al., 2011). They utilized a poly SBS (styrene-*b*-butadiene-*b*-styrene) fiber mat as an elastomeric substrate. Polyelectrolyte gel, Au nanosheets, and P3HT (poly(3-hexylthiophene) nanofibers were employed for gate dielectric, electrodes, and active channel, correspondingly. The comprehensive device structures are presented in Figure 5.2(a)–(c). Au nanosheet electrodes were made through a transfer procedure utilizing P3HT fibers and PDMS pillars were electrospun on the substrate across drain electrodes and source. The transistors were reversibly extended up to a 70% strain ($\varepsilon = 0.7$), as revealed in Figure 5.2(d).

Not merely mechanically, however also electrically, the transistors showed reproducible performance up to 1500 cycles of stretching retrieval at $\varepsilon = 0.7$ (Figure 5.2(e) and (f)). The transistors displayed high hole mobility of 18 cm²/V¨ s and an on/off ratio of 10^5 even under $\varepsilon = 0.7$.

Likened to stretchable transistors, stretchable memories had been less examined. Flexible memories had been developed as a significant component of flexible electronics. For instance, Ouyang et al. (2004) prepared organic resistive memories from a polystyrene film comprising 8-hydroxyquinoline and AuNPs that were inserted amongst two metal electrodes. Ji et al. (2013) made a twistable memory cell series adopting a one diode-one resistor (1D-1R) structure and numerous organic materials for resistor components and diode. Though their memory cell range stably showed high on/off ratios of $>10^3$ up to a bendy angle of 30°, it was cracked at a strain of 2.03%. Lai et al. (2014) made a stretchable organic memory with a bent structure, where a blend of polymer compound and wrinkled graphene bottom electrode was utilized. After a blend of poly(3-butylthiophene (P3BT) and poly(methylmethacrylate (PMMA 3), which worked as the active information storing, was spin-coated on a CVD grownup graphene sheet, the film stack was transmitted on the pre-strained PDMS substrate. This caused a wrinkled organic memory structure, as revealed in Figure 5.3(a). This memory moved from 0 (a low-current state) to 1 (a high-current state) at a threshold voltage of 2.6 V, and the state was firmly maintained even after the exclusion of the applied voltage, which is a usual feature of non-volatile memory (Figure 5.3(b)). As shown in Figure 5.3(c) and (d), the memory behavior was not worsened through a strain up to 50% and the data retention reached 10^4 s (Bhattacharyya et al., 2008; Kim et al., 2011).

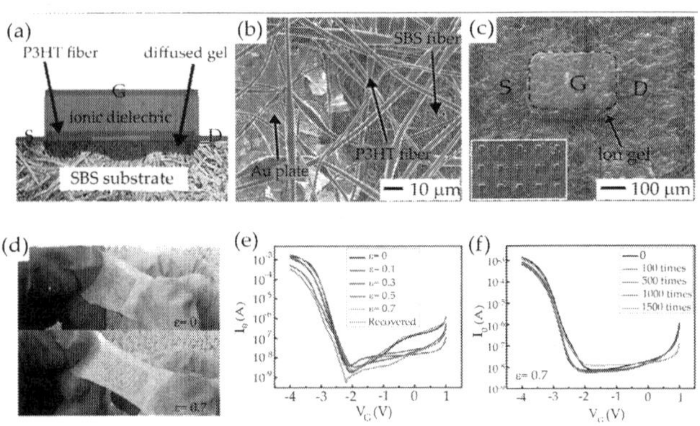

Figure 5.2. (a) Schematic picture; (b) focused SEM image; and (c) low-magnification SEM image of a high stretchable transistor comprising completely of stretchable components; (d) photo images of the stretchable transistor range at two strain states (ε = 0 and 0.7); ID-VG curves relying on (e) the applied strain and (f) the number of stretching release cycles at ε = 0.7. In (a) and (c), D, G, and S represent drain, gate, and source correspondingly.

Source: https://onlinelibrary.wiley.com/doi/abs/10.1002/adma.201400009.

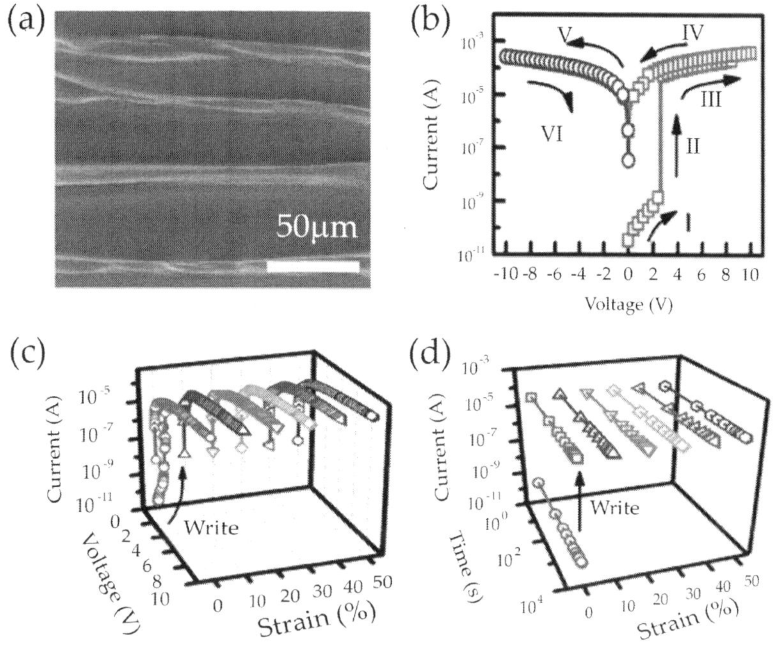

Figure 5.3. (a) Top-view SEM image of the crumpled organic memory; (b) current-voltage characteristics displaying a memory behavior; strain dependent (c) memory switching; and (d) retention time.

Source: https://www.nature.com/articles/am201385.

5.2.3. Stretchable Light-Emitting Diodes

The necessity for stretchable light-emitting systems is ever-growing in fields like rollable lamps, biocompatible light sources, and wearable displays. A common method of implementing stretchable shows is to combine elastic interconnects with organic LEDs or rigid inorganic (Park et al., 2009; Kim et al., 2011). Another method for stretchable displays is to implement intrinsically-stretchable OLEDs (organic LEDs), where all the components are stretchable. Filiatrault et al. (2012) fabricated LEECs (light-emitting electrochemical cells) utilizing stretchable Au/PDMS anodes and stretchable ruthenium (Ru)/PDMS emissive layers. However, they demonstrated huge area emission, the external quantum efficiency (EQE) and the strain tolerance of the devices were comparatively low. Pei's group produced transparent stretchable electrodes formed of fabricated (elastomeric polymer light-emitting device (EPLED) and AgNW-PUA (poly(urethane acrylate)composite, and through merging them with electroluminescent polymer layer (Liang et al., 2013, 2014). The EPLED could release light even at a high strain of 120%. The same group enhanced the performance of transparent stretchable electrodes by introducing graphene oxide (GO) to AgNW percolation networks, as displayed in Figure 5.4(a). The GO soldering turned out to decrease AgNW junction resistance and overwhelm the inter-NW slip. The group fabricated PLED comprised of GO-AgNW-PUA composite electrodes, a polyethyleneimine (PEI) electron transporting layer, and a polymeric emissive layer (Figure 5.4(b)). The PLED could be extended up to 130% (Figure 5.4(c)) and undergo over 100 stretching cycles between 0% and 40% (Xiong et al., 2006; Bokobza, 2007).

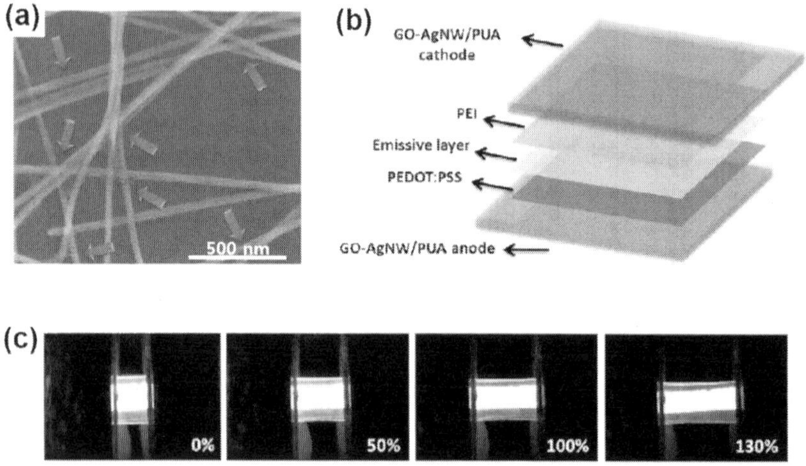

Figure 5.4. (a) SEM image of GO (graphene oxide)-soldered AgNW (silver nanowire) junctions. Red arrows specify GO parts wrapping about AgNW junctions; (b) schematic drawing of a stretchable PLED structure; (c) optical photographs of a PLED (polymer light-emitting device) operating at 14 V under diverse strains.

Abbreviations: PEI: polyethylenimine; PEDOT: poly(3,4-ethylene dioxythiophene); PSS: poly(styrene sulfonic acid); PUA: poly(urethane acrylate).

Source: https://pubs.acs.org/doi/10.1021/nn405887k.

5.2.4. Brief Summary

Representative accomplishments made in the stretchable electronics field are précised in Table 5.1. Of the three applications in this field, stretchable conductors had made the greatest development. They had attained tremendously large stretchability, up to 600%, however maintaining a high electrical conductivity of greater than 10^3 S/cm, which is the level needed for conventional interconnects and electrodes (Brosteaux et al., 2007; Park et al., 2014). However, easier techniques to fabricate such great performance stretchable conductors are required to be developed (Chen et al., 2007; Liu et al., 2015). Stretchable FETs had demonstrated good on/off ratios greater than 10^6; however, their stretchability still required to be further developed for the entirely stretchable applications. Stretchable memories are in their beginning. Stretchable LEDs had made a big development in terms of stretchability; however, their EQE is yet far inferior to their inorganic counterparts (Araby et al., 2014).

Table 5.1. Representative Accomplishments Made in the Field of Stretchable Electronics

Application	Material	Electrical Properties	Mechanical Properties
Stretchable conductors	PU-PPY composites	$\sigma = 10^{-5}$ S/cm	$\varepsilon_b = 160\%$
	Graphene sheets-PU composites	$\sigma = 1.2 \times 10^{-5}$ S/cm	$\varepsilon_b = 374\%$
	PU-PEDOT blends	$\sigma = 10–50$ S/cm	$\varepsilon = 200\%$
	3D PDMS-EGaIn	$\sigma = 24,100$ S/cm	$\varepsilon_b = 220\%$
	Ag flakes-PU composites	$\rho = 2.8 \times 10^{-4}$ Ω¨cm	$\varepsilon_b = 600\%$
Stretchable FETs and memories	SnO_2 NWs/wavy interconnects	On/Off ratio = 10^6	$\varepsilon_b = 40\%$
	P3HT/PS-PCBM/PEN	On/Off ratio > 10^3	$\varepsilon_b = 2.03\%$
	SBS fiber mat/P3HT nanofibers/polyelectrolyte gel	On/Off ratio = 10^5	$\varepsilon = 70\%$
	SWCNT-elastomer composites	On/Off ratio > 10^3	$\varepsilon_b = 70\%$
	PMMA-P3BT/PDMS	Data retention = 10^4 s	$\varepsilon_b = 50\%$

Stretchable LEDs	Ru-PDMS/Au-PDMS	EQE < 1%	ε_b = 27%
	GO-AgNW-PUA composites/PEDOT:PSS/PEI	Current efficiency = 2.0 cd/A	ε_b = 130%
	AuNW-PUA composites	EQE = 4%	ε_b = 120%

Abbreviations: PU: Polyurethane; PCBM: ((6,6)-Phenyl-C_{61}-butyric acid methyl ester); PMMA: Poly(methylmethacrylate); PPY: Polypyrrole; PEN: Polyethylene naphthalate; SWCNT: Single-walled carbon nanotube; P3HT: Poly(3-hexylthiophene); P3BT: Poly(3-butylthiophene).

5.3. STRETCHABLE SENSORS

Though there are an extensive variety of sensors that had been developed to sense a huge number of materials or stimuli, they could be divided into two types: chemical sensors and physical sensors. Chemical sensors identify chemical species in the liquid phase or gas phase, whereas physical sensors notice physical stimuli like strain, temperature, and pressure. Despite this variation in sense targets, the two groups of sensors take benefit of the same sensing principles, which are founded on a variation in the physical properties of a device or sensing material, like capacitance, optical reflectance, and electrical resistance (Mahar et al., 2007). In the sensors field, the demand for stretchable sensors had been rising more and more for exceptional applications like implantable health monitors or body-conformable, electronic skins for intelligent robots and wearable sensory textiles (Thakur, 1988; Perez et al., 2009).

5.3.1. Stretchable Strain Sensors

Strain sensors are devices to exactly detect numerous mechanical deformations like compression, elongation, and bending. In the traditional field of structural health checking, comparatively low stretchability of below 1% would be sufficient for the application. Though for stretchable strain sensors that are the emphasis here, high stretchability is needed in addition to conventional sensor requirements like as good stability, fast response, and high sensitivity. Conventional metal founded strain sensors display the maximum strain of only around 5% and a gauge factor of ~2. Graphene sheets coated on the polymeric substrates had involved great interest, particularly in the aspect of sensitivity (Pandey et al., 2013). Although they had attained a colossal gauge factor of ~10^6, their stretchability was not up to mark (<10%). In comparison, CNT films on elastomeric substrates had shown large stretchability. For instance, Fan et al. (2012) fabricated strain sensors from the CNT networks covered on PU multifilaments and displayed a huge, reversible stretchability of 400%. Gauglitz (2005) fabricated vertically associated SWCNT films on the PDMS substrate and revealed high stretchability of 280% and low creep of 3% at a 100% strain. Regardless of the high stretchability, the measurement factor of the CNT elastomer composites was comparatively low (<5). As another method for recognizing stretchable strain sensors, certain research groups had utilized broken metal film structures on an elastomeric substrate. Puers (1993) formed buckle and microcracks structures in Ti (titanium) films crackled on PDMS through mechanical stretching, and Lacour et al. (2006) made tri-branched microcracks in the Au films electron beam evaporated on the PDMS. Though these strain sensors were simply fabricated and demonstrate strain-dependent resistance alteration, the repeatability of resistance signals required to be further guaranteed (Carone et al., 2002; Faseena et al., 2014).

AgNW-elastomer composites had also been keenly applied for stretchable strain sensors. Zhu and Xu (2012) formed a stretchable capacitive strain sensor by inserting AgNW films on both surfaces of a PDMS sheet. The capacitance of the sensor altered linearly and reproducibly with the functional strain in the range of 0% to 50%. Though, its gauge factor was only ~1. Kim et al. (2013) executed a stretchable resistive strain sensor, having the same approach. PDMS-AgNWs-PDMS sandwich-structured strain sensor (Figure 5.5(a)) presented reproducible resistance alterations under a range of linear strain of 0% to 70% and however under a bending strain (Figure 5.5(b) and (c)). This strain sensor presented nearly no hysteresis up to a strain of 40%, and hysteresis behaviors looked at a 60% strain with occupied resistance recovery at 0 strain. Human motion detection was also confirmed utilizing the strain sensor (Figure 5.5(d)).

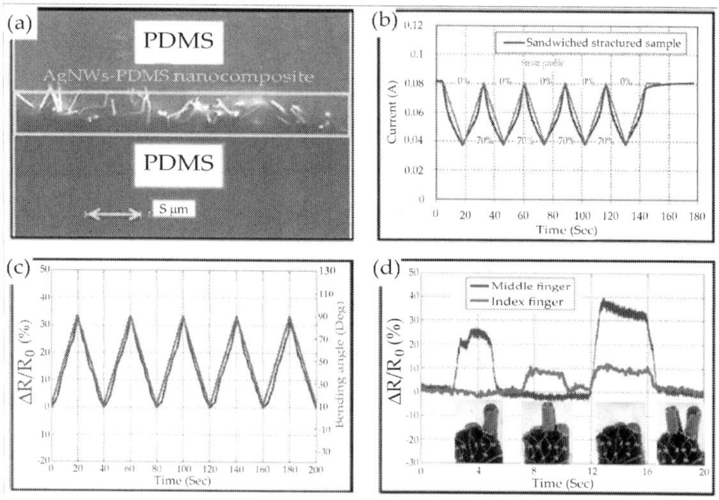

Figure 5.5. (a) Cross-sectional SEM picture of PDMS-AgNWs-PDMS sandwich-structured strain sensor. Repeated replies of the strain sensor (b) to stretching releasing cycles at a strain of 70%; and (c) to bending cycles in the bending angles of 10°–90°; (d) presentation of finger motion detection utilizing the strain sensor.

Source: https://pubs.acs.org/doi/10.1021/nn501204t.

5.3.2. Stretchable Pressure Sensors

Disparate strain sensors, which frequently detect lateral deformations, pressure sensors identify the magnitude of force vertically performing on a unit area of a plane. The main technique for the execution of stretchable pressure sensors is to mix a conductive matter with an elastomer in a precisely devised structure or a composite. In the methods utilizing composite materials, conductive nanostructures like metal nanowires/nanoparticles, CNTs, and CBs, are spread through an insulating elastomer and a resistive alteration under a pressure is measured. Though, the uniform distribution of conductive nanostructures is however to be confirmed more. As a marginally different approach, Liang et al. (2013) covered a PU foam with PPY by soaking the PU foam in a solution comprising naphthalene di-sulfonic acid and pyrrole monomer followed by *in situ* polymerization. They revealed real-time monitoring of the ribcage movement through breathing, utilizing this PPY-coated PU foam (Abbati et al., 2003; Vicentini et al., 2007).

More lately, numerous research groups had developed distinct device structures to attain highly-sensitive pressure sensors. Joo et al. invented a capacitive pressure sensor comprising of a PMMA dielectric layer and an AgNW-embedded PDMS electrode. Regardless of its high sensitivity (>3.8 kPa^{-1}), but, it was not stretchable. Park et al. invented a stretchable pressure sensor that comprised of Au-covered PDMS micropillars (top layer) and PANI nanofibers (bottom layer). This sensor functioned by measuring an alteration of contact resistance amongst the two layers under applied pressure. They attained a high sensitivity of 2.0 kPa^{-1}, a less detection limit of 15 Pa, biaxial stretchability of 15%, a fast response time of 50 ms, from this sensor. As revealed in Figure 5.6(a), the PDMS micro-pyramids were covered with a PEDOT: PSS-PUD (PU dispersion) mixture, which operated as a piezoresistive electrode. Upon the application of pressure, every pyramid inclines to spread laterally, and this rises the contact perimeter (W_{PE}), contact interface area (A_{CI}), and the thickness of the existing path (D_{PE}), thus increasing the current conduction (Figure 5.6(b)). This sensor functioned well even at a high elongation of 40% and presented a very high sensitivity of 10.32 kPa^{-1} at that elongation (Figure 5.6(c)).

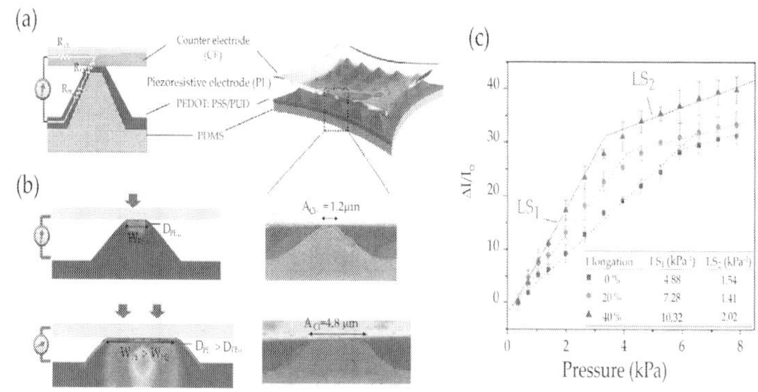

Figure 5.6. (a) Schematic image of a micro-pyramid PDMS array. Discrete PDMS pyramids are covered with a PEDOT: PSS-PUD (Polyurethane dispersion) mixture, which assists as a piezoresistive electrode; (b) finite element analysis data displaying stress distributions and SEM pictures at diverse magnitudes of pressures; (c) relative existing changes relying on the applied pressure, however, the sensor is stretched to a specific elongation. Here, LS1 and LS2 signify the linear sensitivities in particular regions.

Source: https://onlinelibrary.wiley.com/doi/abs/10.1002/adma.201305182.

5.3.3. Stretchable Temperature Sensors

Stretchable temperature sensors had attracted increasing attention, particularly for body-attachable applications. The traditional resistance temperature detector (RTD) takes benefit of the principle that a conductor's resistance alters with temperature. Though, good metals like platinum (Pt) and Au that are usually utilized for temperature sensors don't have stretchability by themselves. To overwhelmed this shortcoming, Chen et al. (2007) utilized a twisting structure of Au and joined it with porous PU. Their temperature sensor displayed good linearity of resistance alteration, good air-permeability, and good stretchability. Likewise, Liu et al. (2015) utilized a coiled structure of graphene that was fixed inside PDMS. This graphene elastomeric offered nonlinear resistance variation with the temperature in the range of 3°C to 100°C and tolerance to strain up to 50%. Thakur (1988) activated a TFTV(thin-film transistor) array with a stretchable PANI nanofiber temperature sensing constituent. The PANI nanofiber sensors and SWCNT TFTs were first prepared on poly(ethylene terephthalate (PET) films, and then they were inserted onto Ecoflex. This temperature sensor showed a high resistance sensitivity of 1.0% per °C, mechanical stability to biaxial strain up to 30%, and a response time of 1.8 s. Moreover, Vicentini et al. (2007) invented an entirely elastomeric gated temperature sensor. For that, a PEDOT: PSS-PUD composite was utilized as a drain, source, and gate material, PU as a gate dielectric, and decrease graphene oxide (R-GO)-PU complex as a temperature sensing channel layer (Figure 5.7(a)). This temperature sensor could expand up to a strain of 70% and showed a high sensitivity of 1.34% and steady response to temperature, however, after 10,000 cycles of stretching at a 30% strain (Figure 5.7(b)).

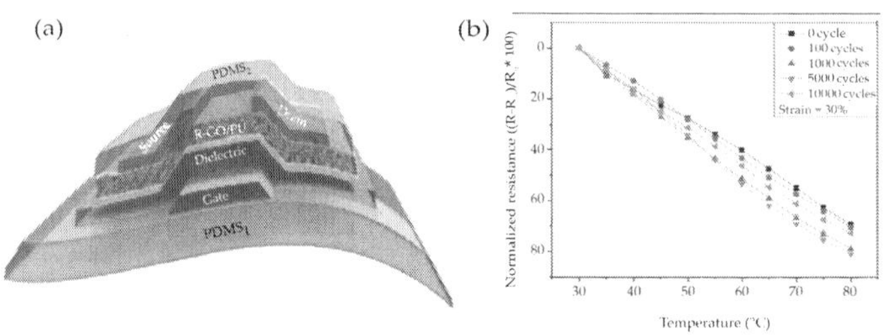

Figure 5.7. (a) Schematic drawing of a whole elastomeric gated temperature sensor; (b) reply of the temperature sensor after cyclic stretching of 0–10,000 cycles at a strain of 30%.

Source: https://onlinelibrary.wiley.com/doi/10.1002/adma.201504441.

5.3.4. Brief Summary

Representative successes made in the field of stretchable sensors are précised in Table 5.2 (Yoshikawa et al., 2006).

Table 5.2. Representative Successes Made in the Field of Stretchable Sensors

Application	Material	Sensing Properties	Mechanical Properties
Stretchable strain sensors	Graphene woven fabric/PDMS	Gauge factor = ~10^6	ε_b = 10%
	CNT networks/PU multi-filament	Gauge factor = ~5	ε_b = 400%
	PDMS/AgNWs/PDMS	Gauge factor = ~5	ε = 0%–70%
	AgNWs/PDMS	Gauge factor = ~1	ε = 0%–50%
Stretchable pressure sensors	PPY-coated PU foam	Sensitivity = 0.0007 mS/N	ε > 1,000%
	AgNW-embedded PDMS/PMMA	Sensitivity > 3.8 kPa^{-1}	Not stretchable
	Micro-pyramid PDMS/PEDOT: PSS-PUD blend	Sensitivity = 10.32 kPa^{-1} Detection limit = 23 Pa	ε > 40%
	Au-coated PDMS micropillars/PANI nanofibers	Sensitivity = 2.0 kPa^{-1} Detection limit = 15 Pa	ε_b (biaxial) = 15%
Stretchable temperature sensors	SWCNT TFT-PANI nanofiber/PET/Ecoflex	Sensitivity = 1.0% per °C	ε_b (biaxial) = 30%
	Graphene embedded in PDMS	Nonlinear *R versus T* in 30–100°C	ε_b = 50%
	PDMS/PEDOT: PSS-PUD composite/PU/(R-GO)-PU composite	Sensitivity = 1.34% per °C	ε_b = 70%

Abbreviations: PANI: Polyaniline; PET: Poly(ethylene terephthalate); TFT: Thin-film transistor.

Altogether three applications had made incredible achievements. Specifically, stretchable strain sensors had displayed great performance in both gauge factor and mechanical stretchability and described their functionality in the area of motion finding. Simple mixtures of elastomers and conductive nanostructures had been confirmed to work efficiently for this application. Stretchable pressure sensors seem to required further development in stretchability, though their sensitivity is sufficiently good. Certain stretchable temperature sensors had fulfilled both requirements of temperature sensitivity and stretchability. Though their material combinations and structures are in general complicated, and further exertions to develop simpler structures would be essential.

5.4. STRETCHABLE ENERGY HARVESTERS

As the global warming problem caused due to the usage of fossil fuel had surged, new energy generation approaches had been in excessive demand. Numerous techniques of harvesting energy from various sources were developed to decrease the consumption of fossil fuel, and few of them had already come into our daily life. Photovoltaic (or solar) cells that transform solar energy into electrical energy are a typical instance. Thermoelectric energy harvesting and piezoelectric energy harvesting had also been enhancing their application areas. Usually, these energy harvesters were formed of inorganic materials like PZT(lead zirconate titanate), Bi_2Te_3 (bismuth telluride), and Si all of which are brittle. Due to the sharply increasing usage of wearable or portable low-power-operated electronics, the necessity for stretchable energy harvesters had been rising larger.

5.4.1. Stretchable Solar Cells

Organic photovoltaic (OPV) cells had been of enhancing interest in numerous applications, mainly for electronic skin, intelligent robotics, and wearable electronics. Around a decade after discovery of record-high power conversion efficiency (PCE) of 2.5% for OPV cells, Mitsubishi Chemical substituted the record with 9.2%. For the conventional bulk heterojunction OPV cells, indium tin oxide (ITO) is utilized as a transparent electrode and it is covered with an hole transport layer (HTL), usually made of PEDOT: PSS. Over the HTL, an active layer that is a blend of acceptor and donor materials is covered by the solution casting method or co-deposition. Till now, the most efficient OPV cells, the PCE of which is above 3.5%, had come from solution-cast P3HT: PCBM ((6,6)-phenyl-C_{61}-butyric acid methyl ester) mixtures. This device structure and material combinations had been maintained as a basic platform for the understanding of stretchable solar cells. A straightforward method of executing a stretchable solar cell is to connect highly effective, however rigid inorganic cells with the stretchable interconnects. Lee et al. (2009) first fabricated GaAs (gallium arsenide) microcells on Si substrate and moved them to the PDMS substrate. Afterward, arc-shaped Au interconnects were transform to bridge the microcells. This microcell array presented a PCE of 13%, a high fill factor of 0.79; however, its stretchability was restricted to below 30%.

Hsu et al. (2011) made stretchable organic solar cells founded on organic materials explained above. They spin-coated P3HT: PCBM layer and PEDOT: PSS layer and placed liquid metal consecutively on pre-strained PDMS substrate. Then, buckles were convinced to the layer stack upon relaxing the PDMS substrate. This buckle structured solar cell displayed endurance to strain up to 22.2%. However, its fill factor (0.38) and PCE (1.2%) were not good. Utilizing a similar material combination for the active layer, an ultrathin OPV cell was invented (Khan et al., 2010). As shown in Figure 5.8(a), the entire thickness of the solar cell is merely 1.9 μm. When this OPV cell fixed on a PET film was attached to a pre-stretched elastomer, it might tolerate a huge compression of up to 80% (Figure 5.8(b)). This solar cell showed enhanced performance with a PCE of 4.2% and a fill factor of 0.61. Though almost reinstated to the actual value after re-stretching it to its early state, the short circuit current of the solar cell reduced with growing compression from 0% to 80% (Figure 5.8(c)).

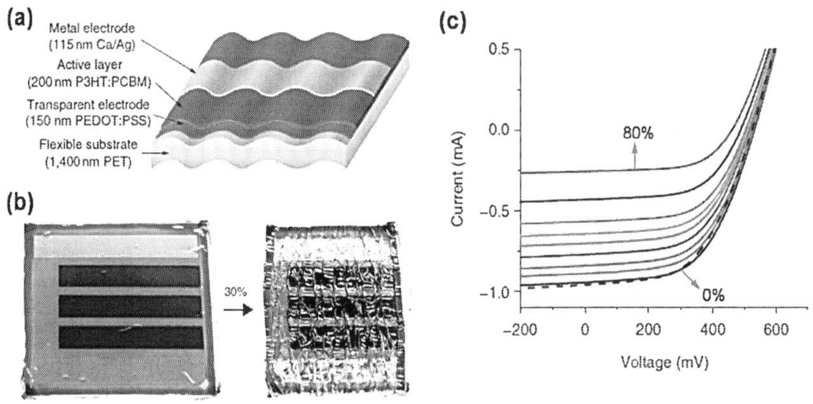

Figure 5.8. (a) Diagram of the ultrathin organic solar cell; (b) stretchable solar cell made simply through attaching the ultrathin cell to a pre-stretched elastomer. It could be re-stretched and compressed; (c) current-voltage curves of the solar cell relying on the compression in the range of 0%–80%. Between the black (0%) and purple (80%) lines, individual colors signify the compression growing with a step of 10%. The black dashed line signifies the device after being restored to its early state.

Source: https://www.nature.com/articles/ncomms1772.

5.4.2. Other Stretchable Energy Harvesters

Piezoelectric energy harvesters transform diverse mechanical stimuli into electricity. Flexible piezoelectric energy harvesters had been stated, which employed numerous nanostructured materials like ZnO nanorods/nanowires, PZT nanowires/ribbons, and polyvinylidene fluoride (PVDF) nanofibers on flexible substrates like paper and polyimide. For instance, El-Tantawy et al. (2009) had got a high-power density of 2.4 μW¨ cm^{-3} utilizing a PZT NW-PDMS nanocomposite. Though these energy harvesters were confirmed to be highly flexible, they needed stretchability. Lee et al. (2009) made a stretchable hybrid nanogenerator (NG) that integrated graphene as the top electrode, poly(vinylidene fluoride-*co*-trifluoro ethylene) as the energy-harvesting layer, and PDMS-CNT compound as the bottom electrode. Their NG might stretch up to a strain of 30%; however, its output power was not huge enough, probably because of the insufficient pyroelectricity and piezoelectricity of the harvesting material. Chen et al. (2013) showed a hyper SEG (stretchable elastic generator) by forming stretchable very long nanowire percolation (VLNP) electrodes on a piezoelectric elastic composite (PEC) comprised of lead magnesium niobate-lead titanate (PMN-PT) particles and MWCNTs discrete in silicone rubber (SR) (Figure 5.9(a)). This SEG displayed a huge, reversible stretchability of ~200% (Figure 5.9(b)) and high-power output of ~4 V and ~500 nA (Figure 5.9(c)).

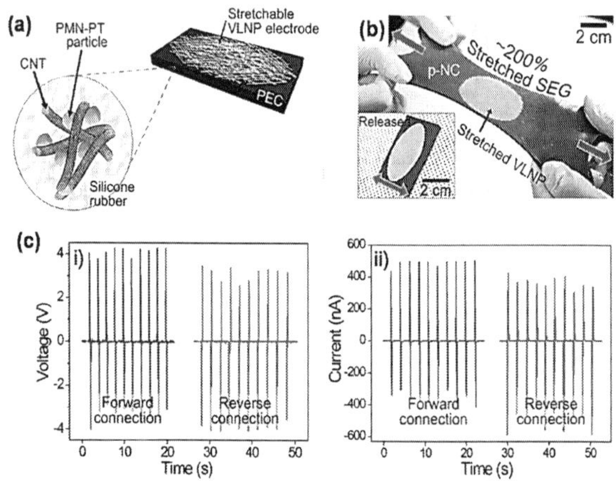

Figure 5.9. (a) Diagram picture of the hyper-stretchable nanocomposite generator (SEG); (b) the SEG could be stretched and free without harm; (c) generated (i) open-circuit voltage and (ii) short-circuit current relying on periodic stretching releasing cycles at a strain of 200%.

Source: https://onlinelibrary.wiley.com/doi/full/10.1002/adma.201500367.

Thermoelectric energy harvesting is a method to produce electrical power from discarded heat. In latest years, organic thermoelectric generators (TEGs) had gained much attention because of numerous advantages over their inorganic counterparts, like as good processibility, flexibility low cost, and low thermal conductivity. The key materials for the organic TEGs were conducting polymers and composites founded on them. Regardless of many successes in flexible organic TEGs, research on stretchable TEGs had been rare. Kim et al. (2008) made a wearable TEG comprising of inorganic thermoelectric components made on glass fabric. Though this TEG fixed in PDMS was flexible, thin, and showed a high output power density of 28 mW·· g^{-1} at a $\Delta T = 50$ K, it was not stretchable. Liang et al. (2013) took the initial step toward a stretchable TEG. They made PPY-SWCNT nanocomposites utilizing an *in situ* oxidative polymerization technique, as depicted in Figure 5.10(a). As a result of the reaction, SWCNTs covered with PPY were got, and their films were prepared through vacuum filtration technique. This nanocomposite film displayed greatly enhanced thermoelectric performance with a power factor of 19.7 μW·· m^{-1}·· K^{-2}, which is the biggest value for PPY composites. Additionally, it could be stretched through 2.6% (Figure 5.10(b)); however, its stretchability still required to be enhanced.

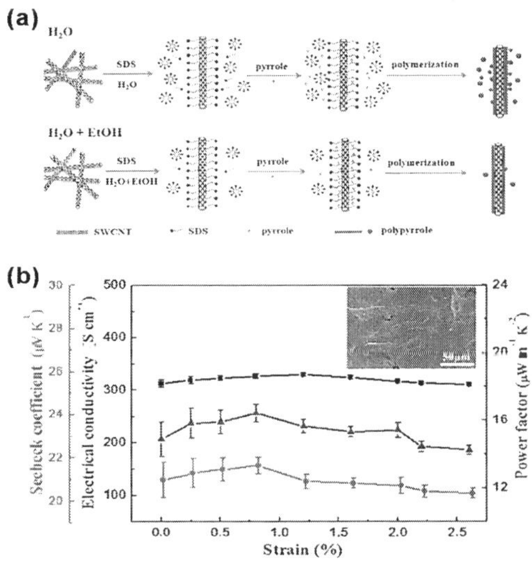

Figure 5.10. (a) Schematic image for the preparation processes of PPY-SWCNT nanocomposites; (b) reliance on the thermoelectric performance of the composite on the mechanical stretching. The inset is an SEM picture of the composite film after a 2.5% stretching.

Source: https://pubs.rsc.org/en/content/articlelanding/2016/TC/c5tc03768a#!divAbstract.

5.5. SUMMARY

Managing with novel technology waves, like intelligent robotics, stretchable/wearable devices, and body-conformable devices, the demand for the conductive elastomers had been whirled. As a reply to this demand, a diversity of research had been intensively started to progress optimal conductive elastomers. Though similar structures and the similar material combination had been employed sometimes in diverse fields of application, comprehensive strategies changed depending on the goal and the target application. In this chapter, research exertions put into three fields of stretchable technology, which are stretchable sensors, stretchable energy harvesters, and stretchable electronics, are reviewed. In every field, how conductive elastomers were integrated into representative devices, exactly how those materials were made, and what performance had been attained with the conductive elastomers are introduced and analyzed. In numerous cases, insulating elastomers like PU, PDMS, and SR were combined either with the conducting polymers like PANI, PPY, and PEDOT: PSS, or with conductive nanostructures like as CNTs AgNWs, and graphene for the execution of conductive elastomers. Further, the conductive elastomers, stretchable functional materials were also integrated to award the main functions of designed devices. Ru/PDMS emissive layer for a stretchable display, PMN-PT particles spread in PDMS for a stretchable piezoelectric energy harvester, and P3HT nanofibers for a stretchable FET, are those instances. PDMS had been most extensively employed as an elastomeric substrate for most applications. As pronounced attainments, a stretchable conductor had shown stretchability of up to 600% and high conductivity of more than 10^3 S/cm. Similarly, a stretchable strain sensor had confirmed stretchability of 400% with a gauge factor of ~5. Though, issues like the development of easy fabrication methods and reliability of conductive elastomers are still to be resolved. Regardless of the notable successes in stretchable applications encouraged by conductive elastomers, it is also factual that numerous stretchable technologies still required to be further explored.

REFERENCES

1. Abbati, G., Carone, E., D'Ilario, L., & Martinelli, A., (2003). Polyurethane-polyaniline conducting graft copolymer with improved mechanical properties. *J. Appl. Polym. Sci., 89*, 2516–2521.
2. Ahn, J. H., & Je, J. H., (2012). Stretchable electronics: Materials, architectures and integrations. *J. Phys. D Appl. Phys., 45*, 103001.
3. Araby, S., Meng, Q., Zhang, L., Kang, H., Majewski, P., Tang, Y., & Ma, J., (2014). Electrically and thermally conductive elastomer/graphene nanocomposites by solution mixing. *Polymer, 55*, 201–210.
4. Araki, T., Nogi, M., Suganuma, K., Kogure, M., & Kirihara, O., (2011). Printable and stretchable conductive wirings comprising silver flakes and elastomers. *IEEE Electron Device Lett., 32*, 1424–1426.
5. Balint, R., Cassidy, N. J., & Cartmell, S. H., (2014). Conductive polymers: Towards a smart biomaterial for tissue engineering. *Acta Biomater., 10*, 2341–2353.
6. Benight, S. J., Wang, C., Tok, J. B. H., & Bao, Z., (2013). Stretchable and self-healing polymers and devices for electronic skin. *Prog. Polym. Sci., 38*, 1961–1977.
7. Bhadra, S., Khastgir, D., Singha, N. K., & Lee, J. H., (2009). Progress in preparation, processing and applications of polyaniline. *Prog. Polym. Sci., 34*, 783–810.
8. Bhattacharyya, S., Sinturel, C., Bahloul, O., Saboungi, M. L., Thomas, S., & Salvetat, J. P., (2008). Improving reinforcement of natural rubber by networking of activated carbon nanotubes. *Carbon, 46*, 1037–1045.
9. Bokobza, L., (2007). Multiwall carbon nanotube elastomeric composites: A review. *Polymer, 48*, 4907–4920.
10. Bokobza, L., Garnaud, G., & Mark, J. E., (2002). Effects of filler particle/elastomer distribution and interaction on composite mechanical properties. *Chem. Mater., 14*, 162–167.
11. Brosteaux, D., Axisa, F., Gonzalez, M., & Vanfleteren, J., (2007). Design and fabrication of elastic interconnections for stretchable electronic circuits. *IEEE Electron Device Lett., 28*, 552–554.
12. Carone, E., D'Ilario, L., & Martinelli, A., (2002). New conducting thermoplastic elastomers. I. Synthesis and chemical characterization. *J. Appl. Polym. Sci., 83*, 857–867.

13. Chen, L., Lu, L., Wu, D., & Chen, G., (2007). Silicone rubber/graphite nanosheet electrically conducting nanocomposite with a low percolation threshold. *Polym. Compos., 28*, 493–498.
14. Chen, M., Tao, T., Zhang, L., Gao, W., & Li, C., (2013). Highly conductive and stretchable polymer composites based on graphene/MWCNT network. *Chem. Commun., 49*, 1612–1614.
15. Chortos, A., & Bao, Z., (2014). Skin-inspired electronic devices. *Mater. Today, 17*, 321–331.
16. Chou, H. H., Nguyen, A., Chortos, A., To, J. W. F., Lu, C., Mei, J., Kurosawa, T., et al., (2015). A chameleon-inspired stretchable electronic skin with interactive color changing controlled by tactile sensing. *Nat. Commun., 6*, 8011.
17. Das, T. K., & Prusty, S., (2012). Review on conducting polymers and their applications. *Polym. Plast. Technol. Eng., 51*, 1487–1500.
18. El-Tantawy, F., Aal, N. A., Al-Ghamdi, A. A., & El-Moss, E. H., (2009). New smart conducting elastomer blends of Bi-based superconductor ceramics nanoparticles reinforced natural rubber/low-density polyethylene for double thermistors, antistatic protectors, and electromagnetic interference shielding effectiveness applications. *Polym. Eng. Sci., 49*, 592–601.
19. Faez, R., Gazotti, W. A., & De Paoli, M. A., (1999). An elastomeric conductor based on polyaniline prepared by mechanical mixing. *Polymer, 40*, 5497–5503.
20. Faez, R., Schuster, R. H., & De Paoli, M. A., (2002). A conductive elastomer based on EPDM and polyaniline II. Effect of the crosslinking method. *Eur. Polym. J., 38*, 2459–2463.
21. Fan, Q., Qin, Z., Gao, S., Wu, Y., Pionteck, J., Mäder, E., & Zhu, M., (2012). The use of a carbon nanotube layer on a polyurethane multifilament substrate for monitoring strains as large as 400%. *Carbon, 50*, 4085–4092.
22. Faseena, N. M., Santhamma, G., Mathew, A. M., & Padmanabhan, P., (2014). Styrene butadiene copolymer-based transparent conducting thin films: Effect of dopants. *J. Elastom. Plast., 46*, 325–337.
23. Filiatrault, H. L., Porteous, G. C., Carmichael, R. S., Davidson, G. J. E., & Carmichael, T. B., (2012). Stretchable light-emitting electrochemical cells using an elastomeric emissive material. *Adv. Mater., 24*, 2673–2678.

24. Fu, Y. P., Weiss, R. A., Gan, P. P., & Bessette, M. D., (1998). Conductive elastomeric foams prepared by *in situ* vapor phase polymerization of pyrrole and copolymerization of pyrrole and N-methyl pyrrole. *Polym. Eng. Sci., 38*, 857–862.
25. Gauglitz, G., (2005). Direct optical sensors: Principles and selected applications. *Anal. Bioanal. Chem., 381*, 141–155.
26. Ge, J., Yao, H. B., Wang, X., Ye, Y. D., Wang, J. L., Wu, Z. Y., Liu, J. W., et al., (2013). Stretchable conductors based on silver nanowires: Improved performance through a binary network design. *Angew. Chem. Int. Ed., 52*, 1654–1659.
27. Ghosh, P., & Chakrabarti, A., (2000). Conducting carbon black filled EPDM vulcanizates: Assessment of dependence of physical and mechanical properties and conducting character on variation of filler loading. *Eur. Polym. J., 36*, 1043–1054.
28. Hansen, T. S., West, K., Hassager, O., & Larsen, N. B., (2007). Highly stretchable and conductive polymer material made from poly(3,4-ethylenedioxythiophene) and polyurethane elastomers. *Adv. Funct. Mater., 17*, 3069–3073.
29. Hilbich, D., Yu, G., Gray, B. L., & Shannon, L., (2015). Characterization of stretchable interconnects fabricated using a low cost metallization transfer process onto PDMS. *ECS J. Solid State Sci. Technol., 4*, S3030–S3033.
30. Hsu, Y. Y., Gonzalez, M., Bossuyt, F., Vanfleteren, J., & De Wolf, I., (2011). Polyimide-enhanced stretchable interconnects: Design, fabrication, and characterization. *IEEE Trans. Electron Devices, 58*, 2680–2688.
31. Ji, Y., Zeigler, D. F., Lee, D. S., Choi, H., Jen, A. K. Y., Ko, H. C., & Kim, T. W., (2013). Flexible and twistable non-volatile memory cell array with all-organic one diode-one resistor architecture. *Nat. Commun., 4*, 2707.
32. Joo, Y., Byun, J., Seong, N., Ha, J., Kim, H., Kim, S., Kim, T., Im, H., Kim, D., & Hong, Y., (2015). Silver nanowire-embedded PDMS with a multiscale structure for a highly sensitive and robust flexible pressure sensor. *Nanoscale, 7*, 6208–6215.
33. Khan, U., May, P., O'Neill, A., & Coleman, J. N., (2010). Development of stiff, strong, yet tough composites by the addition of solvent exfoliated graphene to polyurethane. *Carbon, 48*, 4035–4041.

34. Kim, D. H., & Rogers, J. H., (2008). Stretchable electronics: Materials strategies and devices. *Adv. Mater., 20*, 4887–4892.
35. Kim, D. H., Ahn, J. H., Choi, W. M., Kim, H. S., Kim, T. H., Song, J., Huang, Y. Y., Liu, Z., Lu, C., & Rogers, J. A., (2008). Stretchable and foldable silicon integrated circuits. *Science, 320*, 507–511.
36. Kim, K. H., Vural, M., & Islam, M. F., (2014). Single-walled carbon nanotube aerogel-based elastic conductors. *Adv. Mater., 23*, 2865–2869.
37. Kim, R. H., Bae, M. H., Kim, D. G., Cheng, H., Kim, B. H., Kim, D. H., Li, M., et al., (2011). Stretchable, transparent graphene interconnects for arrays of microscale inorganic light emitting diodes on rubber substrates. *Nano Lett., 11*, 3881–3886.
38. Kim, T. K., Kim, J. K., & Jeong, O. C., (2011). Measurement of nonlinear mechanical properties of PDMS elastomer. *Microelectron. Eng., 88*, 1982–1985.
39. Kim, Y., Zhu, J., Yeom, B., Prima, M. D., Su, X., Kim, J. G., Yoo, S. J., Uher, C., & Kotov, N. A., (2013). Stretchable nanoparticle conductors with self-organized conductive pathways. *Nature, 500*, 59–64.
40. Kojio, K., Furukawa, M., Nonaka, Y., & Nakamura, S., (2010). Control of mechanical properties of thermoplastic polyurethane elastomers by restriction of crystallization of soft segment. *Materials, 3*, 5097–5110.
41. Kuilla, T., Bhadra, S., Yao, D., Kim, N. H., Bose, S., & Lee, J. H., (2010). Recent advances in graphene-based polymer composites. *Prog. Polym. Sci., 35*, 1350–1375.
42. Lacour, S. P., Chan, D., Wagner, S., Li, T., & Suo, Z., (2006). Mechanisms of reversible stretchability of thin metal films on elastomeric substrates. *Appl. Phys. Lett., 88*, 204103.
43. Lacour, S. P., Jones, J., Wagner, S., Li, T., & Suo, Z., (2005). Stretchable interconnects for elastic electronic surfaces. *Proc. IEEE, 93*, 1459–1467.
44. Lacour, S. P., Tsay, C., & Wagner, S., (2004). An elastically stretchable TFT circuit. *IEEE Electron Device Lett., 25*, 792–794.
45. Lai, Y. C., Huang, Y. C., Lin, T. Y., Wang, Y. X., Chang, C. Y., Li, Y., Lin, T. Yet al., (2014). Stretchable organic memory: Toward learnable and digitized stretchable electronic applications. *NPG Asia Mater., 6*, e87.

46. Lang, U., Naujoks, N., & Dual, J., (2009). Mechanical characterization of PEDOT:PSS thin films. *Synth. Met., 159*, 473–479.

47. Lee, J. B., & Khang, D. Y., (2012). Electrical and mechanical characterization of stretchable multi-walled carbon nanotubes/polydimethylsiloxane elastomeric composite conductors. *Compos. Sci. Technol., 72*, 1257–1263.

48. Lee, P., Lee, J., Lee, H., Yeo, J., Hong, S., Nam, K. H., Lee, D., et al., (2012). Highly stretchable and highly conductive metal electrode by very long metal nanowire percolation network. *Adv. Mater., 24*, 3326–3332.

49. Lee, Y. R., Raghu, A. V., Jeong, H. M., & Kim, B. K., (2009). Properties of waterborne polyurethane/functionalized graphene sheet nanocomposites prepared by an *in situ* method. *Macromol. Chem. Phys., 210*, 1247–1254.

50. Liang, J., Li, L., Niu, X., Yu, Z., & Pei, Q., (2013). Elastomeric polymer light-emitting devices and displays. *Nat. Photonics, 7*, 817–824.

51. Liang, J., Li, L., Tong, K., Ren, Z., Hu, W., Niu, X., Chen, Y., & Pei, Q., (2014). Silver nanowire percolation network soldered with graphene oxide at room temperature and its application for fully stretchable polymer light-emitting diodes. *ACS Nano, 8*, 1590–1600.

52. Liu, X., Kuang, W., & Guo, B., (2015). Preparation of rubber/graphene oxide composites with *in-situ* interfacial design. *Polymer, 56*, 553–562.

53. Mahar, B., Laslau, C., Yip, R., & Sun, Y., (2007). Development of carbon nanotube-based sensors: A review. *IEEE Sens. J., 7*, 266–284.

54. Noh, J. S., (2014). Highly conductive and stretchable poly(dimethylsiloxane):poly(3,4-ethylenedioxythiophene): Poly(styrene sulfonic acid) blends for organic interconnects. *RSC Adv., 4*, 1857–1863.

55. Ouyang, J., Chu, C. W., Szmanda, C. R., Ma, L., & Yang, Y., (2004). Programmable polymer thin film and non-volatile memory device. *Nat. Mater., 3*, 918–922.

56. Pandey, S., Goswami, G. K., & Nanda, K. K., (2013). Nanocomposite-based flexible ultrasensitive resistive gas sensor for chemical reactions studies. *Sci. Rep., 3*, 2082.

57. Park, C. W., Jung, S. W., Lim, S. C., Oh, J. Y., Na, B. S., Lee, S. S., Chu, H. Y., & Koo, J. B., (2014). Fabrication of well-controlled wavy

metal interconnect structures on stress-free elastomeric substrates. *Microelectron. Eng., 113*, 55–60.

58. Park, J., Wang, S., Li, M., Ahn, C., Hyun, J. K., Kim, D. S., Kim, D. K., et al., (2012). Three-dimensional nanonetworks for giant stretchability in dielectrics and conductors. *Nat. Commun., 3*, 916.
59. Park, S. I., Xiong, Y., Kim, R. H., Elvikis, P., Meitl, M., Kim, D. H., Wu, J., et al., (2009). Printed assemblies of inorganic light-emitting diodes for deformable and semitransparent displays. *Science, 325*, 977–981.
60. Perez, L. D., Zuluaga, M. A., Kyu, T., Mark, J. E., & Lopez, B. L., (2009). Preparation, characterization, and physical properties of multiwall carbon nanotube/elastomer composites. *Polym. Eng. Sci., 49*, 866–874.
61. Puers, R., (1993). Capacitive sensors: When and how to use them. *Sens. Actuators A, 37*, 93–105.
62. Rogers, J. A., & Huang, Y., (2009). A curvy, stretchy future for electronics. *Proc. Natl Acad. Sci. USA, 106*, 10875–10876.
63. Rogers, J. A., Someya, T., & Huang, Y., (2010). Materials and mechanics for stretchable electronics. *Science, 327*, 1603–1607.
64. Sadhu, S., & Bhowmick, A. K., (2004). Preparation and properties of nanocomposites based on acrylonitrile-butadiene rubber, styrene-butadiene rubber, and polybutadiene rubber. *J. Polym. Sci. B Polym. Phys., 42*, 1573–1585.
65. Saleem, A., Frormann, L., & Soever, A., (2010). Fabrication of extrinsically conductive silicone rubbers with high elasticity and analysis of their mechanical and electrical characteristics. *Polymers, 2*, 200–210.
66. Sekitani, T., Noguchi, Y., Hata, K., Fukushima, T., Aida, T., & Someya, T., (2008). A rubber-like stretchable active matrix using elastic conductors. *Science, 321*, 1468–1472.
67. Shang, S., Zeng, W., & Tao, X., (2011). High stretchable MWNTs/polyurethane conductive nanocomposites. *J. Mater. Chem., 21*, 7274–7280.
68. Shin, G., Yoon, C. H., Bae, M. Y., Kim, Y. C., Hong, S. K., Rogers, J. A., & Ha, J. S., (2011). Stretchable field-effect-transistor array of suspended SnO_2 nanowires. *Small, 7*, 1181–1185.

69. Shin, M. K., Oh, J., Lima, M., Kozlov, M. E., Kim, S. J., & Baughman, R. H., (2010). Elastomeric conductive composites based on carbon nanotube forests. *Adv. Mater., 22*, 2663–2667.

70. Shin, M., Song, J. H., Lim, G. H., Lim, B., Park, J. J., & Jeong, U., (2014). Highly stretchable polymer transistors consisting entirely of stretchable device components. *Adv. Mater., 26*, 3706–3711.

71. Someya, T., Kato, Y., Sekitani, T., Iba, S., Noguchi, Y., Murase, Y., Kawaguchi, H., & Sakurai, T., (2005). Conformable, flexible, large-area networks of pressure and thermal sensors with organic transistor active matrixes. *Proc. Natl. Acad. Sci. USA, 102*, 12321–12325.

72. Stoyanov, H., Kollosche, M., Risse, S., Waché, R., & Kofod, G., (2013). Soft conductive elastomer materials for stretchable electronics and voltage controlled artificial muscles. *Adv. Mater., 25*, 578–583.

73. Sun, Y., Choi, W. M., Jiang, H., Huang, Y. Y., & Rogers, J. A., (2006). Controlled buckling of semiconductor nanoribbons for stretchable electronics. *Nat. Nanotechnol., 1*, 201–207.

74. Tamai, T., (1982). Electrical properties of conductive elastomer as electrical contact material. *IEEE Trans. Compon. Hybrids Manuf. Technol., 5*, 56–61.

75. Thakur, M., (1988). A class of conducting polymers having nonconjugated backbones. *Macromolecules, 21*, 661–664.

76. Vicentini, D. S., Barra, G. M. O., Bertolino, J. R., & Pires, A. T. N., (2007). Polyaniline/thermoplastic polyurethane blends: Preparation and evaluation of electrical conductivity. *Eur. Polym. J., 43*, 4565–4572.

77. Wang, L. X., Li, X. G., & Yang, Y. L., (2001). Preparation, properties and applications of polypyrroles. *React. Funct. Polym., 47*, 125 139.

78. Xiong, J., Zheng, Z., Qin, X., Li, M., Li, H., & Wang, X., (2006). The thermal and mechanical properties of a polyurethane/multiwalled carbon nanotube composite. *Carbon, 44*, 2701–2707.

79. Xu, F., & Zhu, Y., (2012). Highly conductive and stretchable silver nanowire conductors. *Adv. Mater., 24*, 5117–5122.

80. Yoshikawa, H., Hino, T., & Kuramoto, N., (2006). Effect of temperature and moisture on electrical conductivity in polyaniline/polyurethane (PANI/PU) blends. *Synth. Met., 156*, 1187–1193.

CHAPTER 6

Mechanical and Electrical Characteristics of the Elastomeric Nanocomposites

Contents

6.1. Introduction 208

6.2. Basic Problems on Carbon Nanomaterials 209

6.3. Manufacturing Methods of Rubber Nanocomposites 210

6.4. Tensile Properties 211

6.5. Dynamic Mechanical Properties 213

6.6. Electrical Properties Under Strain 217

References 219

6.1. INTRODUCTION

Elastomers that comprise of the polymeric chains with a high degree of mobility and flexibility, show elasticity just like rubber if the polymeric chains are combined into the network structure (Mark and Erman, 1998). Elastomers can suffer large deformations and also have excellent damping features making them suitable for energy dissipation. Rubber-like materials find uses in several sectors ranging from conveyor belts and automobile tires, hoses, aircraft industry, adhesives, etc. The rubber-like elasticity is witnessed above the Tg (glass transition temperature) and above the Tm (melting point) for the crystalline polymers. Natural rubber (NR) experiences strain-tempted crystallization that causes the large upsurge in modulus at very high deformation as the crystallites behave as surplus cross-links in a network (Kraus, 1971; Voet, 1980). From this viewpoint, strain-tempted crystallization can be reflected as the auto-reinforcement of an elastomer. The elastomers that can't undergo strain-tempted crystallization, are normally compounded with additives comprising of fillers such as silica or carbon black (CB) to increase the tensile strength, modulus, and wear resistance of a rubber material (Donnet, 1998; Bokobza, 2004). Though, high levels of loading of the conventional fillers (frequently above 40 phr (parts per 100 parts of rubber by weight) are needed to accomplish the anticipated properties.

Over the last few years, the nanofillers have been broadly utilized in the rubber nanocomposites because of the small size and corresponding upsurge in the surface area permitting substantial enhancement in the properties of a matrix at the low filler loadings. The position of orientation and filler dispersion in a matrix, their aspect ratio, and size along with the collaborations with polymer chains, have been exhibited to be critical parameters that establish the strengthening ability of the nanoparticles.

Iso-dimensional nanofillers (TiO_2 or SiO_2) are the spherical particles that normally can be produced *in situ* through the sol-gel procedure in a polymer. The procedure has been demonstrated to be an efficient and simple approach for the synthesis of the composites where the strengthening phase is ultimately dispersed inside the matrix of polymer which is the basic necessity for accomplishing optimal reinforcement (McCarthy et al., 1998a, b).

Amongst the nanometer-scale strengthening particles, the layered silicates have engrossed a tremendous concern because of the substantial property enhancements that could outcome if a flaked structure is attained with a complete separation of the layers of clay leading to the large interfacial area of contact with a polymer (Bokobza and Diop, 2010; Galimberti, 2011).

The last 10 years have witnessed an increased curiosity for the rod-shaped NFs (nanofillers) and fundamentally for the carbon nanotubes (CNTs). The acknowledgment of their exclusive properties has inspired a huge curiosity in their utilization as advanced fillers in the composite materials (Bokobza, 2007). Particularly, their superior thermal, electrical, and mechanical properties are anticipated to deliver much higher property enhancement as compared to the other nanofillers. For instance, as conductive insertions in the polymeric matrices, CNTs shift the percolation verge to lower loading values as compared to the customary CB particles (Mensah et al., 2015).

Besides CNs, there is a huge interest in graphitic nanostructures including expanded graphite, graphite intercalation compounds, graphite nanoplatelets (GNPs), graphene oxide (GO), and graphene. Their utilization as strengthening fillers for the elastomeric materials holds potential as a specific class of nanocomposites if a layered structure of the graphite, analogous to layered silicate, is typically exfoliated and if the detached nanosheets are dispersed well in a polymeric matrix (Sadasivuni et al., 2014).

The emphasis of this chapter is to outline the state of information in carbon nanostructures, the challenges, properties, and potential uses of the elastomeric matrices packed with carbon nanomaterials (Galimberti et al., 2014a, b).

6.2. BASIC PROBLEMS ON CARBON NANOMATERIALS

Graphite, ample, and therefore cost-effective as the raw material, is made up of stacked parallel 2-D graphene layers comprising of the hexagonal organization of sp^2 carbons. The graphite's layered structure displays a 3-D order in which nearby graphene sheets, normally separated by around 0.337 nanometers, are grasped together by the weak van der Waals forces (Araby et al., 2015; Papageorgiouet al., 2015).

The graphite's layered structure permits intercalation of the chemical species like alkali metals or acids leading to the compounds of graphite intercalation (Chen and Zhao, 2010). Graphite can normally be intercalated by the mixture of nitric acid and sulfuric acid, the first one being utilized as the oxidizing agent. Intercalated graphite can usually be exfoliated or expanded by quick heating creating wormlike or vermicular structures that can typically be succumbed to ultrasonication in order to accomplish graphites with very small thicknesses. GNPs having thicknesses in the range of 2 to 10 nanometers have been attained by re-intercalating with the alkali metal, graphite that has already been intercalated and flaked with the mixture of sulfuric and nitric acid (Viculis et al., 2005). The K-intercalated compound is handled with ethanol (C_2H_5OH) that was discovered to be an efficient exfoliating source. Exfoliation in ethanol (C_2H_5OH) yields hydrogen gas and potassium ethoxide that aids the separation of layers of graphite to form GNPs.

Individual sheets of graphene can be acquired from GO created by chemical oxidation of the graphite trailed by exfoliation. GO bears several oxygen-comprising functional groups. This oxygen (O_2) functionalities have been anticipated to guarantee more compatibility and improved interfacial adhesion with the organic polymers. Unluckily, in silicone/GO composites, the hydrogen bonding interaction amongst GO sheets has been displayed to be stronger as compared to attractive collaboration between the polymer chains and GO (Viculis et al., 2005). Furthermore, Raman spectra of the GO are quite analogous to that of CB therefore displaying that the process of oxidation produces structural defects. The structural defects disturb the interconnected electronic structure of graphite producing the material with low conductivity and unsuitable for the synthesis of the conducting composites. The chemical reduction of GO sheets performed with reducing agents such as hydrazine leads to the partial recovery of conductivity possibly attributed to the restoration of a graphitic network of the sp^2 bonds (Niu et al., 2014).

Wrapping the sheet of graphene into the seamless cylinder triggers the creation of single-walled carbon nanotubes (SWCNTs) whereas multiwall carbon nanotubes (MWCNTs) comprises of several layers of graphene organized in the concentric cylinders having the interlayer distance nearby the distance between the layers of graphene in graphite and usual diameters ranging from 1 to 50 nanometers and lengths varying from micrometers-millimeters and even centimeters (Bokobza et al., 2015).

6.3. MANUFACTURING METHODS OF RUBBER NANOCOMPOSITES

Most of the preparation techniques of nanocomposites are targeted at accomplishing all over the polymer matrix, a uniform and homogeneous filler dispersion. The condition of filler dispersion is known to play the main role in the ultimate properties of a material and to be impacted by the nature of polymer-filler interactions. In addition to fillers, the rubber requires to be compounded with various other additives comprising curing agents, processing aids, coupling agents, and antioxidants. The process of curing that triggers the creation of the crosslinked network might be disturbed by the existence of filler. Solution blending, *in situ* polymerization, and melt mixing is the most usually reported methods in the studies for the preparation of the carbon-centered elastomeric composites (Figure 6.1).

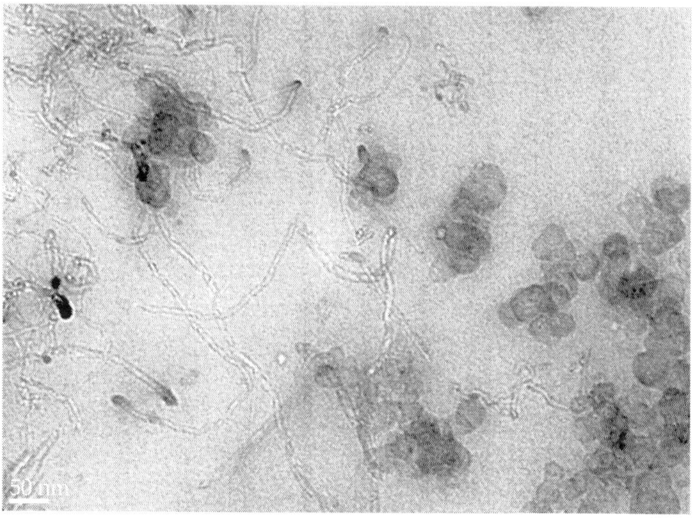

Figure 6.1. Styrene-butadiene rubber (SBR) filled with the double filling (5 phr of MWCNTs + 5 phr of carbon black) (Bokobza et al., 2015).

Source: https://onlinelibrary.wiley.com/doi/full/10.1002/polb.21529.

6.4. TENSILE PROPERTIES

The first result of the integration of the hard filler particles in the soft medium of polymer is an upsurge in elastic modulus. An additional contribution can ascend from polymer-filler collaborations that trigger extra cross-links in a network structure. The upsurge in an efficient extent of cross-linking can normally be assessed by equilibrium swelling using a solvent and using measurements of chain direction that has been displayed to be subtle to a number of the polymer-filler attachments and also to the chemical junctions (Enriquez et al., 2014). In silica-filled poly(dimethylsiloxane) (PDMS) rubbers, for instance, the collaboration between the filler and the polymer is guaranteed with the help of hydrogen bonds amongst the silanols existent on the surface of silica and the atoms of O_2 of PDMS chains. To meet application necessities and therefore impart particular properties to the subsequent material, the interface of polymer-filler can be personalized by utilizing handled silica in which fragments of silanols are deactivated in order to reduce the interactions. The utilization of coupling agents in amalgamation with silica is normally used in non-polar polymers such as hydrocarbon rubbers, to improve the extent of adhesion amongst the filler and the polymer (Bokobza, 2001). Because of their structural characteristics-high particular surface area and high aspect ratio that influences the amount of area with the polymer-CNs are anticipated to report if they are dispersed finely in an elastomeric matrix, substantial improvements in several properties with respect to the conventional fillers.

The hydrocarbon rubbers are normally reinforced by CB and one particular characteristic brought by the active fillers is an upsurge in stress perceived at high deformations. The upsurge in stress at the high elongations is accredited to narrow chain extensibility perceived for the composite due to an upsurge in the efficient extent of cross-linking therefore reducing the obvious molecular weight amongst cross-links. The behavior that reveals the polymer-filler collaborations is to some extent the signature of strengthening. The sudden increase in stress shown by unfilled NR (Figure 6.2(b)) is attributed to the strain-tempted crystallization of the chains of polymer that is a significant characteristic of NR because of its uniform microstructure. The process of crystallization starts at the lower deformation in the existence of CB which might be considered as consequential from the higher chain direction in the orientation of strain permitting the conformational change needed for the creation of crystallites. The chain orientation upsurges with extra cross-links formed by NR-CB interactions. In contrast, the chains of polymer are overstrained by the strain augmentation effects triggered by the insertion of the undeformable filler particles (Wagner, 1976).

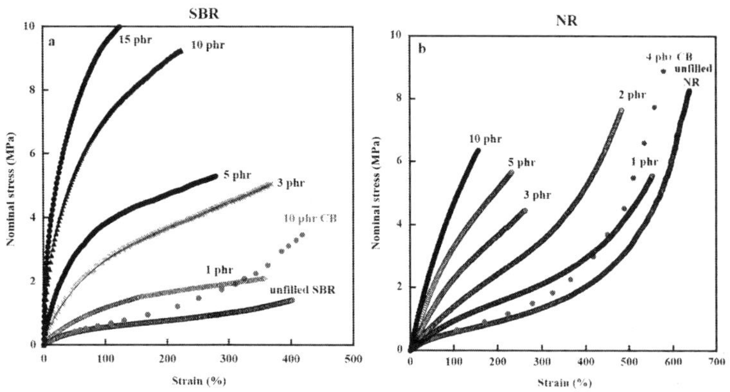

Figure 6.2. *Stress versus strain curves of styrene-butadiene rubber; (a) and natural rubber; (b) composites filled with CB and MWCNTs (in red).*

Source: https://www.mdpi.com/2311-5629/3/2/10.

As observed in Figure 6.2 that shows the stress versus strain curves of NR and SBR composites, considerable enhancement in stiffness, growing with the loading of filler, is reported to matrices by addition of multiwall carbon nanotubes (MWCNTs) with higher levels of strengthening as compared to those offered by the conventional CB particles. However, the CB-filled SBR specimen displays an upsurge in stress at the high strains opposite to the unfilled styrene-butadiene rubber that doesn't display strain-tempted crystallization. The data exhibited in Figure 6.2 were acquired from traditional tensile tests carried out at room temperature on the tensile Instron machine, (model 5565) equipped with a 100 N load cell and the video extensometer. The strips were labeled with two dots with the white marker for recognition by a video extensometer then strained at the strain rate of nearly 0.1/s. The nominal stress, σ, was calculated by the following equation:

$$\sigma = f/A$$

where; f is the elastic force; A is the undeformed cross-sectional area.

6.5. DYNAMIC MECHANICAL PROPERTIES

As mentioned above, the filler-filler collaborations trigger the creation of the filler network that improves the modulus at the low strains and is accountable for strong reduction at the low strains of decreased stress. Generally, the filler network is demonstrated through the strain dependency of storage modulus

G' that displays the non-linear behavior known as the Payne effect (Payne, 1965). G' exhibits the distinguishing amplitude dependence, reducing from the value of modulus at strains upcoming 0, G'_0, to the plateau value $G'\infty$ with a growing amplitude of strain. The amplitude of the Payne effect described as ($G'_0 - G'\infty$) has normally been discovered to be disturbed by the particular surface area, surface treatment, and concentration of the particles of filler along with using temperature. The Payne effect is observably associated with the dispersion state of the particles of filler in a polymer matrix. The utilization of coupling agents in the silica-filled hydrocarbon rubbers, for instance, reduces the amplitude of the Payne effect by discouraging filler-filler collaborations (Bokobza, 2013). In the situation of an ideal dispersion as that acquired by the production of the particles of filler by the process of sol-gel, the Payne effect isn't observed in the entire range of investigated shear strain (Figure 6.3).

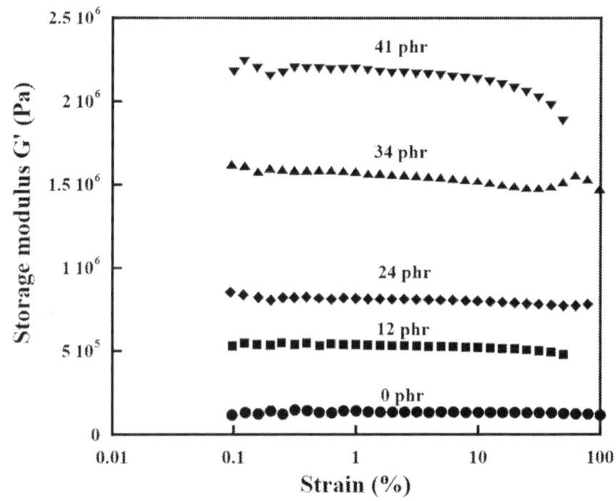

Figure 6.3. Strain dependency of storage modulus of the {poly(dimethylsiloxane)} filled with several amounts of particles of the silica produced in situ by the process of sol-gel.

Source: https://www.mdpi.com/2311-5629/3/2/10.

The decrease in elastic modulus has generally been explained by the breakdown on oscillatory shear, of a filler network created by direct particle interactions or through restrained elastomeric layer where the surface of the particle is recuperated with observing elastomeric molecules (Wang, 1998). However, it must be pointed out that the non-linear behavior can be perceived for the volume fractions of filler lower than the filtration verge, and from this viewpoint, below the filtration verge, the Payne effect can usually be explained by the disruption of filler agglomerated structures (Clément et al., 2005).

The Payne effect has normally been stated for the elastomers filled with diverse carbon materials. Figure 6.4(a) relates the Payne effect in NR composites filled with MWCNTs and CB, and it is displayed that this effect strongly upsurges with the filler amount and with the particles an isometry as it is much significant with MWCNTs as compared with CB at the same filler loading.

Unexpectedly, in a poly(1,4-cis-isoprene) filled with the nano graphite having high shape anisotropy, an observable Payne effect is attained for the content of filler of 40 and 60 phr, which resembles higher volume fractions as compared to those indulged in NR/MWCNTs composites (Figure 6.4(a)) in spite of high aspect ratio of the nano graphite, stated by the authors, anticipated to permit the creation of the filler network at a very low loading of filler. The relative non-linear vibrant viscoelastic response of arranged NR and the composites filled with CNTs alone and together with decreased GO are displayed in Figure 6.4(b). GO was thermally reduced at two diverse temperatures, 200 and 600°C (Ponnamma et al., 2013). The experimental outcomes were fitted to the Goritz and Maier model centered on the filler-rubber collaborations and adsorption procedure of the network chains on filler particles surface (Maier and Göritz, 1996). The reduction of vibrant storage modulus is accredited to the desorption of unsteady bonded chains. It was inferred that the reduction in storage modulus with the shear amplitude specifies an order of the filler-rubber collaboration as NR-CNT > NR –CG600 > NR –CG200. However, it must be figured out that the well-ordered elastomer also exhibits the non-linear behavior opposite to what is anticipated and perceived in Figure 6.4(a) (Kuilla et al., 2010; Wu et al., 2015).

The outcomes published in the studies don't permit to assess of the impact of the nature of the anisotropic particles of carbon on a Payne effect as the non-linear vibrant behavior is strongly associated with the dispersion state. Agglomeration triggers an upsurge in the fraction of volume conforming to the creation of a 3-D network structure that considerably affects the vibrant viscoelastic properties. Here, it is exciting to indicate that in Scotti et al. (2014) work dealing with the shape governed and rod-like nanoparticles having dissimilar aspect ratios in styrene-butadiene rubber, storage modulus at the low strains, G'_0, upsurges with aspect ratio but the values of $G'\infty$

is very analogous for all of the composites filled at the same loading of filler. This gives the meaning that the dissimilarities in strengthening are less efficient once the network breaks which was anticipated since the synthesis of filler has been targeted in order to attain nanoparticles only diverse in shape (Araby et al., 2014; Bokobza et al., 2014).

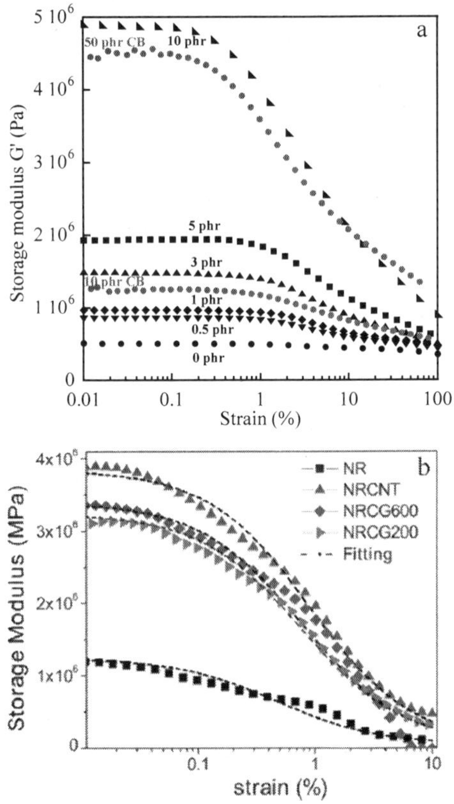

Figure 6.4. Strain dependency of storage modulus: at 25°C of natural rubber composites filled with MWCNTs and CB (a); at 25°C of natural rubber composites filled with carbon nanotubes (5 phr) and with decreased GO (2.5 phr) (CG600 and CG200) as well as with carbon nanotubes (2.5 phr) (b).

Source: https://pubs.rsc.org/en/content/articlelanding/2013/sm/c3sm51978c.

6.6. ELECTRICAL PROPERTIES UNDER STRAIN

Resistivity measurements under strain bring additional understanding into the physical procedures indulged in the deformation of the filled elastomers. The carbon nanomaterials have demonstrated largely their capability to deliver, when integrated into normally insulating media of polymer, electrical conduction at the lower content of filler as compared to CB particles (Ciselli et al., 2010; Singh et al., 2011). At the specified amount of conductive particles known as the percolation verge, a continuous filler network is created across a matrix, and then the material experiences an abrupt change from insulator to the conductor. The percolation verge in several elastomers (SBR, EPDM, and NR) filled with the MWCNTs have been discovered to be nearly 0.5 phr. It has been discovered to be nearly 0.5 phr by Al-Solamy et al. (2012) in the acrylonitrile-butadiene rubber occupied with nanoplatelets of graphite. In the examination of Potts et al. (2011), the stated electrical percolation thresholds accomplished with graphene-centered nanocomposites don't differ considerably from those perceived for Carbon nanotube/polymer composites. Sadasivuni et al. (2014) stated more disseminated values of percolation threshold for graphite and graphene derivative elastomer nanocomposites (Li and Zhong, 2011; Potts et al., 2011).

The parameters of filler like particle size, concentration, structure, orientation, dispersion are the main factors in concluding the composites' electrical properties. Resistivity is also quite sensitive to the mechanical deformation that varies the distribution of the filler. A usual strain dependency of the volume resistivity of the NR/MWCNTs composite is displayed in Figure 6.5. At the content of filler of three phr, the composite is quite above the percolation verge, determined nearly 0.5 phr (Bokobza, 2012). During the first elongating to 100% strain, the resistivity upsurges slowly, therefore, reflecting a reduction in the contacts of filler. This result might be considered as ascending from a configuration of the filler structures in the direction of stretching. White et al. (2009) have offered a 3-D simulation and measurement of electrical conductivity generally above the percolation verge for networks comprising finite, conductive cylinders as the function of axial direction.

Figure 6.5 displays that are unloading slowly the specimen leads to an upsurge in resistivity and just after total elimination of the stress, resistivity is higher as compared to that material which is not deformed therefore displaying that the contacts aren't reformed. The second stretch is completely reversible until the spot where the second and first stretching meets, then the resistivity upsurges again. The 2nd unloading from $\varepsilon = 200\%$ causes a higher resistivity value as compared to that attained in the 1st unloading cycle and the 3rd stretching still shows a decrease, then increase in the resistivity. The distance amongst particles upsurges along the stretch direction but the thickness and width are decreased but upsurge upon unloading till 0 stress causing the loss of contacts amongst conductive particles and the upsurge in resistivity. Re-straining the specimen brings the particles again in the cross-sectional area into interaction therefore clarifying the reversible area of a curve in Figure 6.5 (White et al., 2009).

In the study associated to resistivity behavior of the CB-filled silicone rubber (SR) succumbed to the cyclic loading experiments, Kost et al. (1984) displayed, from stress and comparative experiments of electrical resistivity relaxation that upon the whole retraction of the specimen that follows the step of stretching, the resistivity is much higher as compared to the original value of an unstrained virgin specimen.

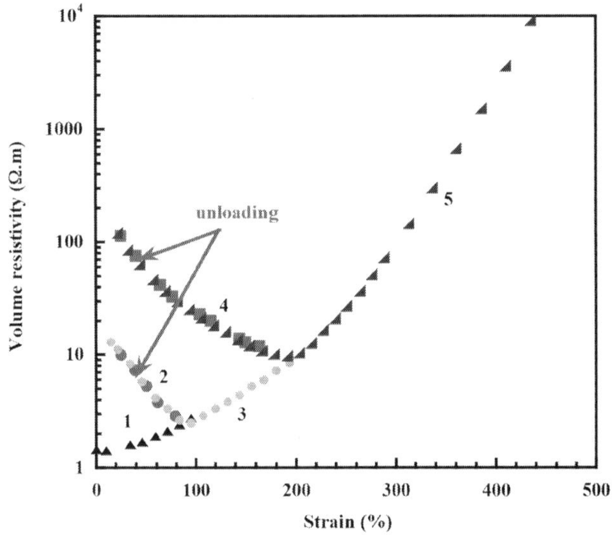

Figure 6.5. Strain dependency of electrical resistivity for the NR filled with three phr of MWCNTs. Dots in red color: resistivity during the process of retraction. The other colors signify the different cycles of loading.

Source: https://www.semanticscholar.org/paper/Mechanical-and-Electrical-Properties-of-Elastomer-Bokobza/30a9ec9b3bfd11f395252154637c682c6e3a2944.

REFERENCES

1. Al-solamy, F. R., Al-Ghamdy, A. A., & Mahmoud, W. E., (2012). Piezoresistive behavior of graphite nanoplatelets based rubber nanocomposites. *Polym. Adv. Technol., 23*, 478–482.
2. Araby, S., Meng, Q., Zhang, L., Kang, H., Majewski, P., Tang, Y., & Ma, J., (2014). Electrically and thermally conductive elastomer/graphene nanocomposites by solution mixing. *Polymer, 55*, 201–210.
3. Araby, S., Meng, Q., Zhang, L., Zaman, I., Majewski, P., & Ma, J., (2015). Elastomeric composites based on carbon nanomaterials. *Nanotechnology, 26*, 112001.
4. Araby, S., Zaman, I., Meng, Q., Kawashima, N., Michelmore, A., Kuan, H. C., Majewski, P., et al., (2013). Melt compounding with graphene to develop functional, high-performance elastomers. *Nanotechnology, 24*, 165601.
5. Bokobza, L., & Diop, A. L., (2010). Reinforcement of silicone rubbers by sol-gel in situ generated filler particles. In: Thomas, S., & Stephen, R., (eds.), *Rubber Nanocomposites: Preparation, Properties, and Applications* (Vol. 1, pp. 63–85). John Wiley & Sons (Asia): Jin Xing Distripark, Singapore, Chapter 3.
6. Bokobza, L., (2001). Filled elastomers: A new approach based on measurements of chain orientation. *Polymer, 42*, 5415–5423.
7. Bokobza, L., (2004). The reinforcement of elastomeric networks by fillers. *Macromol. Mater. Eng., 289*, 607–621.
8. Bokobza, L., (2007). Multiwall carbon nanotube elastomeric composites: A review. *Polymer, 48*, 4907–4920.
9. Bokobza, L., (2012). Enhanced electrical and mechanical properties of multiwall carbon nanotube rubber composites. *Polym. Adv. Technol., 23*, 1543–1549.
10. Bokobza, L., (2013). Elastomeric composites based on nanospherical particles and carbon nanotubes: A comparative study. *Rubber Chem. Technol., 86*, 423–448.
11. Bokobza, L., Bruneel, J. L., & Couzi, M., (2015). Raman Spectra of carbon-based materials (from graphite to carbon black) and of some silicone composites. *C J. Carbon Res., 1*, 77.
12. Bokobza, L., Pflock, T., Lindemann, A., Kwiryn, D., & Dos, S. C. P., (2014). Thermal conductivity and mechanical properties of composites

based on multiwall carbon nanotubes and styrene-butadiene rubber. *Kautsch. Gummi Kunststoffe, 67*, 45–50.

13. Chen, G., & Zhao, W., (2010). Rubber/graphite nanocomposites. In: Thomas, S., & Stephen, R., (eds.), *Rubber Nanocomposites: Preparation, Properties, and Applications* (Vol. 1, pp. 527–550). John Wiley & Sons (Asia): Jin Xing Distripark, Singapore. Chapter 19.
14. Ciselli, P., Lu, L., Busfield, J. J. C., & Peijs, T., (2010). Piezoresistive polymer composites based on EPDM and MWNTs for strain sensing applications. *E-Polymers*.
15. Clément, F., Bokobza, L., & Monnerie, L., (2005). Investigation of the Payne effect and its temperature dependence on silica-filled polydimethylsiloxane networks. Part I: Experimental results. *Rubber Chem. Technol., 78*, 211–231.
16. Donnet, J. B., (1998). Black and white fillers and tire compound. *Rubber Chem. Technol., 71*, 323–341.
17. Enriquez, E., De Frutos, J., Fernandez, J. F., & De La Rubia, M. A., (2014). Conductive coatings with low carbon-black content by adding carbon nanofibers. *Compos. Sci. Technol., 93*, 9–16.
18. Galimberti, M., (2011). *Rubber Clay Nanocomposites: Science, Technology, and Applications* (Vol. 1, pp. 1–44). John Wiley & Sons: Hoboken, NJ, USA.
19. Galimberti, M., Cipoletti, V., Musto, S., Cioppa, S., Peli, G., Mauro, M., Gaetano, G., et al., (2014b). Recent advancements in rubber nanocomposites. *Rubber Chem. Technol., 87*, 417–442.
20. Galimberti, M., Kumar, V., Coombs, M., Cipoletti, V., Agnelli, S., Pandini, S., & Conzatti, L., (2014a). Filler networking of a nanographite with a high shape anisotropy and synergism with carbon black in poly(1,4-cis-isoprene)-based nanocomposites. *Rubber Chem. Technol., 87*, 197–218.
21. Kost, J., Narkis, M., & Foux, A., (1984). Resistivity behavior of carbon-black-filled silicone rubber in cyclic loading experiments. *J. Appl. Polym. Sci., 29*, 3937–3946.
22. Kraus, G., (1971). Reinforcement of elastomers by carbon black. *Adv. Polym. Sci., 8*, 155–237.
23. Kuilla, T., Bhadra, S., Yao, D., Kim, N. H., Bose, S., & Lee, J. H., (2010). Recent advances in graphene-based polymer composites. *Prog. Polym. Sci., 35*, 1350–1375.

24. Li, B., & Zhong, W. H., (2011). Review on polymer/graphite nanoplatelets nanocomposites. *J. Mater. Sci., 46*, 5595–5614.

25. Maier, P. G., & Göritz, D., (1996). Molecular interpretation of the Payne effect. *Kautsch. Gummi Kunststoffe, 49*, 18–21.

26. Mark, J. E., & Erman, B., (1998). *Rubberlike Elasticity, A Molecular Primer* (Vol. 1, pp. 1–33). John Wiley & Sons: Hoboken, NJ, USA.

27. McCarthy, D. W., Mark, J. E., & Schaeffer, D. W., (1998a). Synthesis, structure, and properties of hybrid organic-inorganic composites based on polysiloxanes. I. Poly(dimethylsiloxane) elastomers containing silica. *J. Polym. Sci. Part B Polym. Phys., 36*, 1167–1189.

28. McCarthy, D. W., Mark, J. E., Clarson, S. J., & Schaeffer, D. W., (1998b). Synthesis, structure, and properties of hybrid organic-inorganic composites based on polysiloxanes. II. Comparisons between poly(methyl phenyl siloxane) and poly(dimethylsiloxane), and between titania and silica. *J. Polym. Sci. Part B Polym. Phys., 36*, 1191–1200.

29. Mensah, B., Kim, H. G., Lee, J. H., Arepalli, S., & Nah, C., (2015). Carbon nanotube-reinforced elastomeric nanocomposites: A review. *Int. J. Smart Nano Mater., 6*, 211–238.

30. Niu, R., Gong, J., Xu, D., Tang, T., & Sun, Z. Y., (2014). Influence of molecular weight of polymer matrix on the structure and rheological properties of graphene oxide/polydimethylsiloxane composites. *Polymer, 55*, 5445–5453.

31. Papageorgiou, D. G., Kinloch, I. A., & Young, R. J., (2015). Graphene/elastomer nanocomposites. *Carbon, 95*, 460–484.

32. Payne, A. R., (1962). The dynamic properties of carbon black-loaded natural rubber vulcanizates. Part I. *J. Appl. Polym. Sci., 6*, 57–63.

33. Payne, A. R., (1965). Dynamic properties of natural rubber containing heat-treated carbon blacks. *J. Appl. Polym. Sci., 9*, 3245–3254.

34. Ponnamma, D., Sadasivuni, K. K., Strankowski, M., Guo, Q., & Thomas, S., (2013). Synergistic effect of multi-walled carbon nanotubes and reduced graphene oxides in natural rubber for sensing application. *Soft Matter, 9*, 10343–10353.

35. Potts, J. R., Dreyer, D. R., Bielawski, C. W., & Ruoff, R. S., (2011). Graphene-based polymer nanocomposites. *Polymer, 52*, 5–25.

36. Sadasivuni, K. K., Ponnamma, D., Thomas, S., & Grohens, Y., (2014). Evolution from graphite to graphene elastomer composites. *Prog. Polym. Sci., 39*, 749–780.

37. Scotti, R., Conzatti, L., D'Arienzo, M., Di Credico, B., Giannini, L., Hanel, T., Stagnaro, P., et al., (2014). Shape controlled spherical (0D) and rod-like (1D) silica nanoparticle morphology on the filler reinforcing effect. *Polymer, 55*, 1497–1506.
38. Singh, V., Joung, D., Zhai, L., Das, S., Khondaker, S. I., & Seal, S., (2011). Graphene-based materials: Past, present and future. *Prog. Mater. Sci., 56*, 1178–1271.
39. Viculis, L. M., Mack, J. J., Mayer, O. M., Hahn, H. T., & Kaner, R. B., (2005). Intercalation and exfoliation routes to graphite nanoplatelets. *J. Mater. Chem., 15*, 974–978.
40. Voet, A., (1980). Reinforcement of elastomers by fillers: Review of period 1967–1976. *J. Polym. Sci. Macromol. Rev., 15*, 327–373.
41. Wagner, M. P., (1976). Reinforcing silicas and silicates. *Rubber Chem. Technol., 49*, 703–774.
42. Wang, M. J., (1998). Effect of polymer-filler and filler-filler interactions on dynamics properties of filled vulcanizates. *Rubber Chem. Technol., 71*, 520–589.
43. White, S. I., Di Donna, B. A., Mu, M., Lubensky, T. C., & Winey, K. I., (2009). Simulations and electrical conductivity of percolated networks of finite rods with various degrees of axial alignment. *Phys. Rev. B, 79*, 024301.
44. Wu, X., Lin, T. F., Tang, Z. H., Guo, B. C., & Huang, G. S., (2015). Natural rubber/graphene oxide composites: Effect of sheet size on mechanical properties and strain-induced crystallization behavior. *Exp. Polym. Lett., 9*, 672–685.

CHAPTER 7

Applications of Elastomers in Defense and Aerospace

Contents

7.1. Introduction.......224

7.2. Acoustic Applications.......224

7.3. Aircraft Tires.......230

7.4. Airships.......234

7.5. Inflatable Seacraft.......237

References.......240

7.1. INTRODUCTION

Nine months back to Pearl Harbor, the United States Navy presents an article that explained the use of rubber, stated: 'Throughout the extensive network of Naval ships and bases, rubber is playing a vital part in the nation's first line of defense' (Wexler, 2018). Certainly, over the subsequent 60+ years, the usage of elastomers in the military has increased to a large extent. Its applications range from the sublime (i.e., a 900 kg rubber disk used for the ejection of torpedoes) to the mundane (i.e., ersatz rubber bricks used for covering sensors while Marine reconnaissance missions). Potential future and current uses of rubber are reviewed in this chapter for aerospace and navy applications (Choi et al., 1994; Carlson et al., 2009). For several decades, the space program and the military both have fostered the improvement of novel technologies, and it is a fact for elastomers. Lifetime is a primary interest, Navy ships with a 30-year life cycle (having aircraft shippers constructed for 50 years of lifetime); however, the applications discussed here were/are generally advanced technologies. Even though particulars of military applications are confidential sometimes, and a more general obstacle to the news is the possessive nature of the materials. Yet manufacturing is not done by the US government, private companies supply the rubber items and are in charge of much of their manufacturing. This confined the explanation herein to mostly a qualitative nature.

7.2. ACOUSTIC APPLICATIONS

In several acoustic applications, rubber is very generally used particularly by the Navy, which takes the support of the acoustic impedance match within water and rubber. If there is the same acoustic impedance of two materials, then there will be no reflections at their interface, where the acoustic impedance is defined as the product of the sound speed and the mass density of material (Hagelberg and Corsaro, 1985). The sound speed is proportional to the modulus (shear modulus for shear waves, or bulk modulus for longitudinal waves) and the square root of the ratio of the density for low loss materials. Because the bulk modulus differs weakly within elastomers, fine-tuning the acoustic impedance of rubber depends chiefly on regulating its density. The acoustic characteristics of different rubbers of interest for the navy are present, even though particular formulations contribute to be proprietary (Capps, 1989). A measure of the decrease in intensity of the transmitted wave is referred to as the attenuation coefficient of rubber, the sound amplitude losing exponentially with the product of the attenuation coefficient and the distance traveled. The attenuation coefficient is proportional to the proportion of bulk loss modulus and the bulk storage modulus for longitudinal waves (that are oscillating in the line of the sound propagation). The loss tangent for longitudinal strains is normally below 10^{-3} for elastomers. Hence, sound waves can be moved over long distances with the slightest distortion (Ferry, 1980).

If the aim is to avoid detection, then sound waves should be attenuated. This can be achieved by transforming the longitudinal sound waves into shear waves (i.e., 'mode conversion'), in the consideration of the order of unity of loss tangent for sheared rubber (Sperling, 1990). There are several methods to achieve this mode conversion, e.g., by involving additions such as gas bubbles or small glass spheres, or by restraining the rubber as a thin film within two rigid surfaces. The interfacial rubber in such enclosed geometry impairs in a shear (or extensional) mode, that is quickly attenuated.

The rubber itself can be prepared to be extensively destructive at the frequencies of concern. Maximum energy dissipation takes place on the visco-elastic response of the material commence at the rubber-glass transition region at the applied temperature and frequency. The transition occurs exactly above the traditional glass transition temperature (T_g) for high-frequency sound waves. As evaluated with scanning calorimetry at usual heating rates, T_g corresponds to a malformation time scale of around 100 seconds. Comparably high T_g elastomers are needed to acquire a rubber-glass transition at room-temperature and acoustic frequencies, because of the very large sufficient activation energy for local motion in polymers (a change of 10°C temperature can modify the relaxation time to three orders of magnitude (Wetton, 1978; Roland et al., 2001).

7.2.1. Sonar Rubber Domes

Sonar Rubber Domes also facilitates underwater 'vision' to sea vessels, even though sound permits their detection. On Navy surface ships, the sonic transducers are shielded with rubber (see Figure 7.1) and hold on a steel-reinforced rubber dome. On smaller vessels, the dome is located on the keel (where keel means the bottom beam operating from bow to stern), whereas on large ships, it is situated on the bow (that is, the forward part of the lower hull). The objective of the keel and bow domes is to give a hydrodynamically smooth surface, to secure the transducer, and to lessen noise from the water flow (Traissac et al., 1995; Mott et al., 2002). In October 2000, the recent was demonstrated in the terrorist bombing of the *USS Cole*. Regardless of the loss to the ship itself, the dome and its transducer ride out undamaged (see Figure 7.2). The sonar dome should communicate with minimal distortion in the sound energy, and cannot be interrupted by the flow of seawater. Initially, the domes were manufactured with steel, but they had fragile sound transmission, were naive to marine fouling (from slime-producing bacteria, seaweed, barnacles, etc.), and corrosion, and demanded internal supports, which interfered with the sound. In 1965, the first rubber sonar dome was deployed, where its actual production starts in 1972 (Corsaro and Sperling, 1990).

Figure 7.1. Sonar rubber bow dome.

Source: *http://polymerphysics.net/pdf/BookChapter%20RubTechHandbook.pdf.*

Figure 7.2. After a terrorist attack in Yemen in October 2000, the guided-missile vandal USS Cole back to the United States.

Source: http://polymerphysics.net/pdf/BookChapter%20RubTechHandbook.pdf.

Regardless of the scrambling on the mounting of the ship on a salvage transport vessel, also the 150 m^2 gash in the port side of the hull (refers to Figure 7.2), the bow dome was still operational (seen in Figure 7.3 hanging off the edge of the transport). After fixing that involved restoration of 550 tons of external steel plating, the Cole mounted back to the sea duty after 18 months (Roland et al., 2001; Nemery et al., 2002).

The globally largest molded rubber articles are the sonar bow domes (see Figure 7.3). Their weight is 8,600 kg, is 6.4 m wide, 11 m long, and almost 2.5 m in height. The rubber wall thickness differs up to the utmost 20 cm. The manufacturing entails manual lay-up of multiple steel-cord reinforced polychloroprene plies. Structural rigidity is provided by the steel cords. The spacing of the cords should be less than the wavelength of sound (for example, 1.5 m at 1 kHz) to evade intrusion with an acoustic performance. The rubber itself has less absorption at the sonar frequencies. The dome is vulcanized in an abundant autoclave and fabricated on an open (one-sided) mold. With nearly 95,000 liters of water, the bow dome is magnified to an internal pressure of 240 kPa. Its locale which is down the baseline of the ship decreases the hydrodynamic resistance (Roland and Lee, 1990; Roland, 2009).

Figure 7.3. [Left image]: Bow of the Zr. Ms. Mercuur (A900); [right image] interior sonar dome structure.

Source: https://www.semanticscholar.org/paper/Structural-Response-of-Bow-Structures-with-Sonar-Velner/cfc45221e11177928090d6650777fcbeb7698854/figure/2.

7.2.2. Active Sonar

The use of active sonar to detect the surface ships and submarines is another acoustic utilization of rubber. The usage of inherent quiet diesel-electric submarines and the developments in quieting of naval vessels involves greater detection capacity as compared to passive sonar (Figure 7.4) (Tyler, 1992; Moradi et al., 2009).

Figure 7.4. Composite keel dome consisting of a soft rubber core and two resin-impregnated fiberglass outer layers. The geometry of the system is adjusted to be acoustically transparent over a restricted range of frequency.

Source: https://www.polymerphysics.net › RubberChemTech_77_542_04.

Since wavelength is in inverse relation with the attenuation coefficient for sound waves so generally, active sonar operates on low frequencies (100–500 Hz). That is why it has the ability to detect over several kilometers. As the actual range affects the absorption and dispersion of sound waves produced by the water, because it relies on the sea state, size of the target, depth, and as well as on temperature. This is obvious that resolution is decreased by the longer wavelength and hence it would be difficult to discriminate among undersea objects (Hastings et al., 1996).

7.2.3. Insulation

Sound is similar to acoustic applications, and is a form of mechanical energy, for vibration damping, elastomers are used, e.g., in motor mounts and shock. All through, thermal insulation and acoustic use the ship rubber. For instance, for the prevention of moisture condensation, foam rubber is positioned on submarine walls whereas, for insulation, PU rubber blocks are placed on the sides of the ship reactor. Decoupling tiles of rubber are placed at the back of the ship bulkhead for the modesty of shipboard machinery. NR sandwiched is used within polyethylene (PE) rings and steel by the thrust reducer on submarines (see Figure 7.5), which decouples the hull against propeller vibrations. In the propeller shafts belonging to the V-22 Osprey tilt-rotor aircraft, elastomeric bearings are utilized (Crum and Mao, 1996).

Figure 7.5. Thrust reducer utilized for decoupling the submarine propeller shaft. The natural rubber is intervening between an outer layer of polyethylene and the steel ring.

Source: https://www.dshseals.com/product/stainless-steel-d-type-triple-rotary-lip-ptfe-oil-seal.

7.3. AIRCRAFT TIRES

Aircraft tires and automotive passenger tires are different in several aspects. Almost 85% of aircraft tires come about bias ply instead of radial design (Li et al., 2002; Davidson et al., 2005). About a century back, when the earliest bias aircraft tires were manufactured, to just before 1981, airplanes didn't use radial tires. Aircraft tires are principally based on NR, as the high temperatures gained during landing and takeoff (beyond 100°C). Tires of massive aircraft are often filled with nitrogen (Figures 7.6 and 7.7) (White et al., 2007).

Figure 7.6. Military automobile with polyurea-coated steel plates on back and sides of the truck and sides of HHMWV.

Source: https://www.stripes.com/news/troops-add-improvised-armor-to-humvees-1.16236.

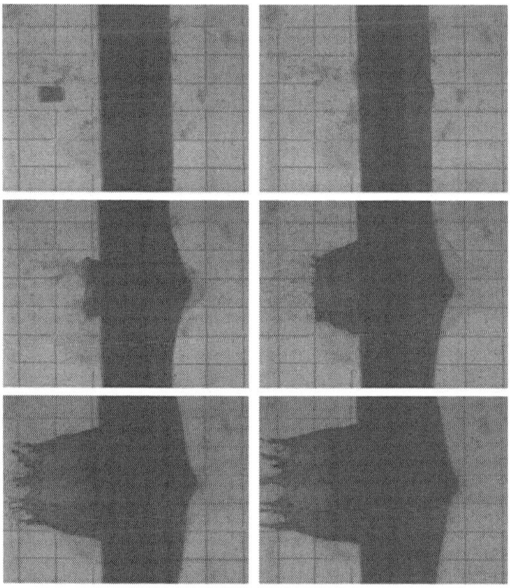

Figure 7.7. Fragment-simulating 0.50 caliber projectile striking steel armor plate covered with 1,4-polybutadiene rubber. A projectile that is shot with a rifled Mann barrel and traveling at 3 km/h speed makes first interaction rubber-

coated side with a dent and does not penetrate through the high-hard steel plate. The grid lines are separated 25 mm apart.

Source: http://polymerphysics.net/pdf/RubberChemTech_77_542_04.pdf

Even though military aircraft tires do the same operation to their commercial counterparts, the requirements enforced on the previous in time can be more severe. The speed rated for the tires of fighter jets, e.g., the F/A-18 and F-16 are over 400 km/h. The inflation pressures can be as strong as 2 MPa. Obviously, fighter aircraft are faster and smaller than bombers. Normally, jet fighters have two primary tires, a single- or double-tire at its nose and the recent being steerable. The primary tires are considerably large in size, e.g., on the F/A-18, nose tires are 56 x 16.8 cm *versus* 76 x 29 cm (these figures refer to the section width and diameter of the tire, respectively). The tire size is implied by the heaviness of the aircraft. If the gross weight of the F-16 was raised by about a factor of two, a considerably larger tire size had to be taken up (Veirh, 1992; Klose, 2010).

Normally, tire wear occurs because of frictional forces, coming out from sliding at the road-tread interface. In automobile tires, the main reason for treadwear is a lesser extent braking, and cornering manoeuvers (Rogers et al., 1986). The alike situation occurs for aircraft while using land-based runways – around 70% of tread wear is imputed to braking. Although, much of this wear does not occur during the interaction of the tire with the runway. Wear is strongly dependent on the alignment of the tire in the direction of the airplane. Crosswinds hinder the tires from easily rolling on the runway. The tire is critically distorted out of the contact patch (on massive aircraft, the extent of landing speeds can be 425 km\hour along with an acceleration of 20 G). While taxiing accounts for 90–95% of the total tread wear, there was an exertion of the merger of the steering forces and crosswinds on the rollout. Generally, the tires of a bomber (e.g., B-52) are changed, depending on the condition, usually after every 30 to 40 flights. At the time of current combat operations in Iraq, 2 to 3 tires of a B-52 get replaced every day. Each tire stands 1.2 m high and weighs 230 kg (Modzelewski and Allcock, 2014).

The situation for carrier-based fighter aircraft is not so good, because the tire treads become worn after just 2 or 3 flights. Ironically, this minimal life period has not much to do with the instant deceleration of carrier landings or the severe acceleration of launching, for example, a F/A-18 Hornet a speed of 265 km/h is attained in a space of 75 m while launching (Figure 7.8) (Tappin, 994).

Figure 7.8. Military crew mounts a tire of a B-52H Stratofortress. The changed tire is drained after its removal. The replacement process takes approximately 20 minutes.

Source: https://world-defense.com/threads/us-armed-forces.3753/page-24.

The acceleration depends on steam-assisted catapulting, and the landings use arresting wires instead of depending on tire friction. Hence, participation in treadwear is less for both situations. A further factor is that landings and carrier launches occur when the ship changes direction into the wind, which subtracts (or adds) to the relative speed, putting a minimal expectation on the tires (Figure 7.9) (Boutilier, 2017).

Figure 7.9. The world's largest warship, Nimitz-class aircraft carrier. Maneuvering needed to locate planes for launching is subjected to much of the tread wear.

Source: https://www.pinterest.com/pin/717127940645263175/.

7.4. AIRSHIPS

There are two kinds of airships. One of them is blimps that are non-rigid airships inadequate to any internal support structure. They depend on the pressure of the internal gas, confined in a rubber-coated fabric, to provide lift and support. Normally, the gas is helium at pressures higher than ambient. Hydrogen offers around 10% more tendency to float (lifting capacity approximate to 1.1 kg/m^3 *vs* 1.02 kg/m^3 for He) but is still limited to use despite its flammability (its energy content is threefold higher as compared to gasoline). Rigid airships (for example, 'Zeppelins') have rigid internal frames and are generally out-of-date (Hewitt et al., 2010). An envelope is the main element of an airship which enclose the pressurizing gas. PCR impregnated polyester fabric was an envelope on the Goodyear blimps which were utilized by the military by the time of World War II. Gas leaks are very gradual due to the low pressure and the immense gas volume (around 7 million liters). Therefore, mechanical durability and manufacturing ease, as well as the need for low permeability guide the elastomer selection. In the experiments of the British Ministry of Defense in 1994, hundreds of bullets punctured an airship envelope, and the craft remained aloft and inflated for many hours (Gander et al., 2006).

During the last century, the blimps were employed by the US military on the North American coasts as 'radar pickets' to catch incoming aircraft and for enemy submarine reconnaissance. For watching surfacing submarines, they facilitate being convoy escorts as well. During World War II, from ~89,000 ships that were carried by blimps, none of them were wasted by enemy fire. In 1962, these blimps were demilitarized and became obsolete because of the development of satellites. From 1980, tethered blimps ('Tethered Aerostat Radar System') has been used by the Air Force that is 60 m long. They were used for watching low-flying aircraft of smugglers by flying at 4.5 km above the US-Mexican border (Mata Pacheco, 2014; Windahl et al., 2014).

In 2002, the development of 'High Altitude Airship' was begun by the US Department of Defense that could act as a communication relay-an alternate option to low Earth orbit communication satellites, they are supposed to be unmanned and are solar-powered blimp for reconnaissance (using both radar and cameras) to identify low-flying missiles. The blimp would be out of the range of ground-launched anti-aircraft weapons when flying at the peak of 20 km. Prototypes will have a volume of 150 million liters (i.e., 25 times greater than the Goodyear blimp), 150 m long, and lift an 1800 kg payload. One technical barrier is accomplishing enough buoyancy in the stratosphere-new technologies might be required for the envelope. By the reason that the internal gas will spread on the blimp growth, hence it will be filled with air initially, which is dismissed by pressurized helium gas when higher altitudes are achieved. Initially, flight experiments are pondered in 2008 (Figure 7.10) (Puskas et al., 2005, 2009).

Figure 7.10. Trial of various Naval airships (in circa 1930), including a kite balloon (upper craft showing in the figure at left), 5 spherical balloons, two J-class non-rigid blimps in the center, the USS Los Angeles in the middle distance, and a ZMC-2, a rigid metal-skinned blimp (showing at right).

Source: https://commons.wikimedia.org/wiki/File:Usn-airships_(cropped).jpg.

More recently, a research program was introduced by the US Department of Defense for the development of huge airships ('WALRUS') for swift strategic airlifts. The goal is to ferry tremendous payloads of transport cargo and armed forces without refueling over very protracted distances (e.g., 10,000 km). By 2008, contemporary schemes demand the development of a 30-ton prototype. The air-buoyant blimp should have to be weather-independent and need just minimal ground assistance to meet the aim of implementing troops globally within a couple of days (Venus, 1983; Katz, 1984).

The 'space blimp' is another more determined application of airship technology which is a mile-long airship engaged to ferry payloads to and from low Earth orbit (with height 350–1400 km). The space blimp would have very thin walls, to be very light, and with a high surface to figure ratio.

Therefore, it could not endure atmospheric pressures; hence, a traditional airship should be used to carry the payload up to ~35 km raised to the Earth's surface (which is the highest height attainable with traditional air buoyancy) (Figures 7.11 and 7.12).

Figure 7.11. Artist's vision of huge blimp WALRUS proposed for swift strategic airlifts.

Source: http://www.oocities.org/usarmyaviationdigest/airborneaircraftcarriers.htm.

Figure 7.12. Prototype space blimp being float tested, 53 m long.

Source: www.jpaerospace.com.

7.5. INFLATABLE SEACRAFT

7.5.1. Combat Rubber Raiding Craft

Combat Rubber Raiding Craft (CRRC) is a crucial military use of rubber which is a 4.6 m long boat constructed of a polychloroprene and aramid-reinforced blend of chlorosulfonated polyethylene (CSPE). Navy SEALs (Sea-Air-Land Team) and Marines use the CRRC to hit the shoreline while dropped in littoral waters. Usually, CRRC is preloaded with weapons, fuel, outboard motors, and other equipment, then let down from a helicopter when fully inflated. Then a cargo hook holds it in place, as troops slide using a rope. This whole operation can be done in less than a minute. The CRRC can also be commenced from the air by leaping on top of a wooden platform using an attached parachute and be deck-launched or subsurface from submarines. Sometimes known as the Zodiac, in reference to the developer, the CRRC is used by civilian fire, law enforcement, and rescue authorities as well (Figure 7.13) (Hoppie, 1982; Mott et al., 2003).

Figure 7.13. Combat rubber raiding craft carrying military personnel between the shore and ship.

Source: https://commons.wikimedia.org/wiki/File:Combat_Rubber_Raiding_Craft_(CRRC)_operations_from_USS_Green_Bay_150712-N-NI474-294.jpg.

7.5.2. Hovercraft

A continuation of the rubber inflatable idea is the military hovercraft, called 'boat that flies' or Landing Craft Air Cushion (LCAC). US Navy first deploy it in 1987, and the armed services of many countries currently use these hovercraft. LCAC is an aquatic automobile that uses two 3 MW (4000 HP) gas turbine engines in order to operate four centrifugal fans. The fans provide a hoist, permitting the craft to ride on an air cushion above the water or ground. No unit of the hull passes through the water. This enormously minimizes friction and permits the vehicle to move independently of ground, permissive transport over sand, rock, water (free of tides), and so on, and to remove barriers up to 1.2 meters high. LCAC can approach exceedingly 80% of the earth's coastlines. Another similar pair of engines drive the craft forward with speeds up to 50 knots (93 km/h). A hovercraft can lift a payload of more than 50 tons (45,000 kg), also with 30 personnel. LCAC has a 19,000-liter fuel tank and it devours approximately 3,800 liters of fuel per hour-this allows a range of around 400 km fully loaded. Their stopping distance surpasses 500 m (Gelling et al., 1988; Roland, 2004).).

Elastomers are used in two major scopes of hovercraft: A NR/Nylon fabric produces a skirt, which covers from the sides in the shape of conical flaps to manage the downward airflow. Moreover, a polychloroprene/Nylon fabric makes an inflated tube with the periphery. This serves being an auxiliary floatation device and to cushion the craft (Figure 7.14) (Roberts, 1988; Little, 1999).

Figure 7.14. Landing craft air cushion with variable-pitch propellers, and propelled by two four-bladed 3.7 m diameter. The polychloroprene/nylon tube can be viewed at the water interface.

Source: https://www.businessinsider.com/video-of-lcac-marine-hovercraft-2017-7.

The vital military use of hovercraft is shuttling equipment and troops, for the ship to sea transport, and for traversing beaches. In shallow waters, they are exceptionally effective for minesweeping and have also been utilized for humanitarian alleviation. LCAC perform a significant part in the latter 'Iraqi Freedom' military campaign (Barton, 1998; Kim et al., 2003).

REFERENCES

1. Barton, R., (1998). Just before nature: The purposes of science and the purposes of popularization in some English popular science journals of the 1860s. *Annals of Science, 55*(1), 1–33.
2. Boutilier, J. A., (2017). The great game goes to sea: Growing maritime insecurity in the Indo-Pacific region. *The Future of Sea Power*, 23, 1–33.
3. Capps, R. N., (1989). *Elastomeric Materials for Acoustical Applications* (Vol. 1, pp. 1–33). Naval research lab Orlando Fl underwater sound reference detachment.
4. Carlson, C. A., Morris, R., Parsons, R., Treusch, A. H., Giovannoni, S. J., & Vergin, K., (2009). Seasonal dynamics of SAR11 populations in the euphotic and mesopelagic zones of the northwestern Sargasso sea. *The ISME Journal, 3*(3), 283–295.
5. Choi, I. S., Roland, C. M., & Bissonnette, L. C., (1994). An elastomeric ejection system. *Rubber Chemistry and Technology, 67*(5), 892–903.
6. Corsaro, R. D., & Sperling, L. H., (1990). *Sound and Vibration Damping with Polymers* (Vol. 1, pp. 1–24). American Chemical Society.
7. Crum, L. A., & Mao, Y., (1996). Acoustically enhanced bubble growth at low frequencies and its implications for human diver and marine mammal safety. *The Journal of the Acoustical Society of America, 99*(5), 2898–2907.
8. Davidson, J. S., Fisher, J. W., Hammons, M. I., Porter, J. R., & Dinan, R. J., (2005). Failure mechanisms of polymer-reinforced concrete masonry walls subjected to blast. *Journal of Structural Engineering, 131*(8), 1194–1205.
9. Ferry, J. D., (1980). *Viscoelastic Properties of Polymers* (Vol. 1, pp. 1–24). John Wiley & Sons.
10. Gander, P. H., Marshall, N. S., James, I., & Le Quesne, L., (2006). Investigating driver fatigue in truck crashes: Trial of a systematic methodology. *Transportation Research Part F: Traffic Psychology and Behavior, 9*(1), 65–76.
11. Gelling, I. R., Porter, M., & Roberts, A. D., (1988). In: Roberts, A. D., (ed.), *Natural Rubber Science and Technology* (Vol. 1, pp. 1–22). Oxford University Press, New York.

12. Hagelberg, M. P., & Corsaro, R. D., (1985). A small pressurized vessel for measuring the acoustic properties of materials. *The Journal of the Acoustical Society of America, 77*(3), 1222–1228.

13. Hastings, M. C., Popper, A. N., Finneran, J. J., & Lanford, P. J., (1996). Effects of low-frequency underwater sound on hair cells of the inner ear and lateral line of the teleost fish *Astronotus ocellatus*. *The Journal of the Acoustical Society of America, 99*(3), 1759–1766.

14. Hewitt, C. N., Lee, J. D., Mackenzie, R., Barkley, M. P., Carslaw, N., Carver, G. D., & Yin, X., (2010). Overview: Oxidant and particle photochemical processes above a south-east Asian tropical rainforest (the OP3 project): introduction, rationale, location characteristics and tools. *Atmospheric Chemistry and Physics, 10*(1), 169–199.

15. Hoppie, L. O., (1982). The use of elastomers in regenerative braking systems. *Rubber Chemistry and Technology, 55*(1), 219–232.

16. Katz, H., (1984). *U.S. Patent No. 4,423,829* (Vol. 1, pp. 1–33). Washington, DC: U.S. Patent and Trademark Office.

17. Kim, T. W., Park, S. S., Setoguchi, T., & Takao, M., (2003). The effect of rotor geometry on the performance of a Wells turbine for wave energy conversion (Part II: The suitable choice of blade design factors). *Journal of the Korean Solar Energy Society, 23*(3), 55–61.

18. Klose, C. D., (2010). Human-triggered earthquakes and their impacts on human security. *Nature Preceding, 1*, 1–13.

19. Li, M., Johnson, C. P., Adams, M. B., & Sarna, S. K., (2002). Cholinergic and nitrergic regulation of *in vivo* giant migrating contractions in rat colon. *American Journal of Physiology-Gastrointestinal and Liver Physiology, 283*(3), G544–G552.

20. Little, C., (1999). Customer futures. *Scenario and Strategy Planning, 1*, 12 14.

21. Mata, P. O. C., (2014). *Cenozoic Structure, Stratigraphy, and Paleogeography of the Lower Magdalena Basin* (Vol. 1, pp. 1–32). Colombia.

22. Modzelewski, T., & Allcock, H. R., (2014). An unusual polymer architecture for the generation of elastomeric properties in fluorinated polyphosphazenes. *Macromolecules, 47*(19), 6776–6782.

23. Moradi, L. G., Davidson, J. S., & Dinan, R. J., (2009). Response of bonded membrane retrofit concrete masonry walls to dynamic pressure. *Journal of Performance of Constructed Facilities, 23*(2), 72–80.

24. Mott, P. H., Roland, C. M., & Corsaro, R. D., (2002). Acoustic and dynamic mechanical properties of a polyurethane rubber. *The Journal of the Acoustical Society of America, 111*(4), 1782–1790.
25. Mott, P. H., Roland, C. M., & Hassan, S. E., (2003). Strains in an inflated rubber sheet. *Rubber Chemistry and Technology, 76*(2), 326–333.
26. Nemery, B., Fischler, B., Boogaerts, M., Lison, D., & Willems, J., (2002). The Coca-Cola incident in Belgium, June 1999. *Food and Chemical Toxicology, 40*(11), 1657–1667.
27. Puskas, J. E., Ebied, A., Lamperd, B., & Kumar, B., (2009). *U.S. Patent No. 7,614,349* (Vol. 1, pp. 1–23). Washington, DC: U.S. Patent and Trademark Office.
28. Puskas, J. E., Kumar, B., Ebied, A., Lamperd, B., Kaszas, G., Sandler, J., & Altstädt, V., (2005). Comparison of the performance of vulcanized rubbers and elastomer/TPE/iron composites for less-lethal ammunition applications. *Polymer Engineering and Science, 45*(7), 966–975.
29. Roberts, A. D., (1988). *Natural Rubber Science and Technology* (Vol. 1, pp. 1–32). Oxford University Press.
30. Rogers, W. P., Armstrong, N. A., Acheson, D. C., Covert, E. E., Feynman, R. P., Hotz, R. B., & Sutter, J. F., (1986). *Report of the Presidential Commission on the Space Shuttle Challenger Accident, 1*, 1–33.
31. Roland, C. M., & Lee, G. F., (1990). Interaggregate interaction in filled rubber. *Rubber Chemistry and Technology, 63*(4), 554–566.
32. Roland, C. M., (2004). Naval applications of elastomers. *Rubber Chemistry and Technology, 77*(3), 542–551.
33. Roland, C. M., (2009). Naval and space applications of rubber. In: *Chapter 5 in Rubber Technologists Handbook* (Vol. 1, pp. 1–34). Smithers Rapra Technology Limited Shawbury (UK).
34. Roland, C. M., Ngai, K. L., Santangelo, P. G., Qiu, X. H., Ediger, M. D., & Plazek, D. J., (2001). Temperature dependence of segmental and terminal relaxation in atactic polypropylene melts. *Macromolecules, 34*(18), 6159–6160.
35. Sperling, L. H., (1990). *Sound and Vibration Damping with Polymers: Basic Viscoelastic Definitions and Concepts, 1*, 1–23.
36. Tappin, L., (1994). Analyzing data relating to the challenger disaster. *The Mathematics Teacher, 87*(6), 423.

37. Traissac, Y., Ninous, J., Neviere, R., & Pouyet, J., (1995). Mechanical behavior of a solid composite propellant during motor ignition. *Rubber Chemistry and Technology, 68*(1), 146–157.
38. Tyler, G. D., (1992). The emergence of low-frequency active acoustics as a critical antisubmarine warfare technology. *Johns Hopkins APL Technical Digest, 13*(1), 145–159.
39. Veirh, A. G., (1992). A review of important factors affecting treadwear. *Rubber Chemistry and Technology, 65*(3), 601–659. 330696+9.
40. Venus, Jr. F., (1983). *U.S. Patent No. 4,387,833* (Vol. 1, pp. 1–30). Washington, DC: U.S. Patent and Trademark Office.
41. Wetton, R. E., (1978). Design and measurement of polymeric materials for vibration absorption and control. *Applied Acoustics, 11*(2), 77–97.
42. Wexler, R., (2018). Life, liberty, and trade secrets: Intellectual property in the criminal justice system. *Stan. L. Rev., 70*, 1343.
43. White, P. R., Leighton, T. G., Saunders, K., & Jepson, P., (2007). *Review of the Mechanisms by Which Anthropogenic Noise May Cause Cetacean Stranding, 1*, 1–33.
44. Windahl, J., Kihlgren, E., Liedström, E., & James, I., (2014). To maintain older family members lifestyle and self-identity for a meaningful daily life in nursing homes and home care: The relatives' perspective. *Clin. Nurs. Stud., 2*, 129–142.

CHAPTER

8 Bio-Medical Uses of Elastomeric Materials

Contents

8.1. Introduction 246

8.2. Developments on Elastomeric Blends and Composites for Medical Applications 247

8.3. Biostable and Biocompatible Materials 249

8.4. Modern Elastomer's Superior Mechanical Properties 249

8.5. Intelligent Bio-Materials 250

8.6. Overview of Products Based on Elastomer Biomedical Applications 250

8.7. Thermoplastic Composites and Elastomers 251

References 259

8.1. INTRODUCTION

For biomedical applications, a prominent increase in the usage of composites and elastomeric blends has been observed in recent years. They are being widely used in medical devices, chronic implants such as vascular grafts and heart valves, and other utilities such as cardiac assist pumps (Gogolewski, 1991; Christenson et al., 2005). These materials have superior biocompatibility, and they offer good mechanical properties as well; this provides them an edge over ceramics and metals as biomaterials, and due to this, the use of these materials has been promoted. Moreover, elastomeric blends (such as thermoplastic elastomers (TPEs)) are cost-effective, can be easily recycled, and their degree of purity is high (extractable compounds having low level). Excellent barrier properties are exhibited by TPE blends (Gunatillake et al., 2003; Ripple and Simons, 2007). They can be easily tailored and in this way, we can use them for achieving desired material properties. The tailoring property of these materials is another factor that supports their use in such applications (Lelah and Cooper, 1986; Szycher, 1998).

A material chosen for implants and medical devices should possess good mechanical properties, biostability, and biocompatibility. By the term biocompatibility, we mean that the material being used should not cause severe allergic reactions or any harm to individuals or patients. Simultaneously, the biological environment (such as bio durability and biostability), where it is being used, should not adversely affect its chemical, physical, and mechanical properties. Furthermore, the material should have such mechanical properties that can resist the predominant loading conditions. The implant or medical device should withstand these conditions without any sort of premature failure (Szycher and Reed, 1992; Stokes McVenes, 1995).

Though usually composites and elastomeric blends have most of their usage in cardiovascular-related applications, they are also utilized in many dental and reconstructive medical applications. These include knee replacements and hip-joint where for complimenting traditional materials (i.e., ceramics and metals), elastomeric blends are being used, cosmetic surgery (for example, breast implants like related applications), skincare, dental implants, and medical devices such as renal dialyzers, intraocular lenses, and contact lenses (Onatea et al., 2001; Khorasani et al., 2005). For biomaterials, the worldwide demand estimated for biomaterials, in the year 2000, was over 300 billion dollars/annum. This represents almost 7 to 8%

of total healthcare spending across the world. Each year, this demand grows significantly, rendering the annual growth rate to be 10% (Brandon et al., 2001; Koo, 2008). The approximation of medical devices consumed per year is shown in Table 8.1, and it is categorized in accordance with their area of application. For biomedical applications, the advances in the elastomeric blends and composite's development have improved enough that they have gained consumer confidence. As a result, the demand for these materials has increased. For replacing soft and hard tissues, the demand for improved biomaterials lost due to age-related degeneration, trauma, or disease will always be enough. In this chapter, we will be discussing the biomedical applications of composites and elastomeric blends (Barr and Bayat, 2009; Barr et al., 2010).

Table 8.1. World's Annual Biomaterials Consumption (Approximated)

Area of Application	**Number of Devices Used/Year (Approximated)**
Contact lens	75,000,000
Stent (cardiovascular)	2,000,000
Intraocular lens	7,000,000
Renal dialyzer	25,000,000
Hip and knee prostheses	1,000,000
Catheter	300,000,000
Vascular graft	400,000
Heart valve	200,000
Breast implants	300,000
Dental implants	500,000

Source: Ratner (2007).

8.2. DEVELOPMENTS ON ELASTOMERIC BLENDS AND COMPOSITES FOR MEDICAL APPLICATIONS

Under the light of medical applications, we can group the materials used in them, into three categories: natural polymers (such as biomolecular materials), synthetic polymers, and inorganic materials (alloys, ceramics, glasses, and metals). Both, elastomeric composites and blends are characterized under synthetic polymers. They are being widely used in medical applications such as hip-joint prostheses and vascular grafts (Lysaght and O'Loughlin, 2000;

Shanshan et al., 2010). For medical applications, the need for materials having good biodegradability and biostability, biocompatibility, and mechanical properties has primarily driven the development of elastomeric composites and blends. For these applications, natural biopolymers can serve as the best materials, as these materials are both resorbable and bioactive. Natural biopolymers are subject to regulatory issues, processing variability, tedious purification techniques, and when used as allografts or xenografts, they can potentially induce dangerous immune responses (Bettinger, 2011).

This resulted in exploring synthetic polymers so that they can be used in medical applications. A unique set of properties is possessed by synthetic elastomers. These properties set them highly beneficial for biomedical applications. In contrast to inorganic materials, these polymers have superior biocompatibility and are due to this fact, they are widely used in various medical applications. The risk of device failure is minimized by their good mechanical properties. For synthetic polymer chemists, the chemical composition of polymers or elastomers offers substantial opportunities for tailoring the structures to the point where they meet specification requirements. The desired requirements are achieved by altering the combination of the monomers and by changing the processing conditions. For instance, in contrast to low-density polyethylene (PE), the tensile strength can be 100 times more for high molecular weight PE (Kurtz Steven, 2009). We can also modify the elastomeric composites and blends by using secondary processes such as grafting/crosslinking, blending, or filling in a manner that is cost-effective so that it suits the application.

In medical applications, the synthetic polymers that are currently used include thermoset elastomers, i.e., Xylitol, and Poly(glycerol-co-sebacate) based elastomers, and TPEs, i.e., PE and polyurethanes (PUs). They have some processing limitations prohibiting their applications in few clinical applications. For instance, their curing conditions require vacuum environments and high temperatures (more than 100 C). Hence, in most of the applications, TPEs have benefits and are usually favored. Recently, the developments on elastomers (including their composites and blends) for medical applications have targeted to achieve superior mechanical properties and excellent biostability and biocompatibility. This ensures that the risk of device failure is reduced and in patients, it reduces allergic reactions. It also minimizes device rejection risk.

8.3. BIOSTABLE AND BIOCOMPATIBLE MATERIALS

The biostability and biocompatibility of implants and medical devices are essential if the devices need to operate appropriately under a specific application while having an appropriate response from the host. In the recent decade, both these properties have become elementary in the manufacture of implants and medical devices. The aim is to improve the service life and performance of implanted medical devices and to minimize adverse allergic reactions. Because of this, latex-based device usage was discontinued, as in some patients, these products were resulting in severe allergic reactions (Tilak, 2001). In most applications, the latex-based elastomers are replaced by silicone elastomers and PU. These applications include heart valves and vascular grafts. Good biostability and biocompatibility are offered by PU and silicone elastomers (Kanyanta and Ivankovic, 2010). In the past years, researches have helped in reaching this achievement (Colas and Curtis, 2009). A lot is yet to come and research is continued. Medical implant's service life can be significantly shortened by poor biostability, resulting in undesirability. Recently, researches regarding the Kang et al. (2010) work on PU blends has significantly helped in attaining better biostability, resulting in longer service life: this is achieved by elastomer blending. A PU elastomer was produced by Kang et al. (2010), it exhibits minor (almost negligible) degradation in its properties: once it has been exposed to cobalt chloride/hydrogen peroxide solution for 14 weeks (Kanyanta, 2009).

8.4. MODERN ELASTOMER'S SUPERIOR MECHANICAL PROPERTIES

The physical and mechanical properties necessary for the good performance of its physiological function should be possessed by each medical device and biomaterial. For instance, a tendon material should be flexible and strong, a heart valve leaflet should be tough and flexible, contrary to them, an artificial cartilage substitute should be elastomeric (it is the ability to achieve larger deformations without getting rupture) and soft. Similarly, good mechanical durability should be possessed by the medical device. For example, flex-fatigue life is necessary for heart valve leaflets (mechanical ones), these leaflets have to flex about 70 times in one minute, and it should not tear for the patient's lifetime (at least for 10 years). It is unfortunate that at the same time, only some elastomeric blends possess all three of them together, i.e., good mechanical properties and excellent biostability and biocompatibility. Silicone and PU elastomers are two elastomers that meet these requirements.

Generally, over their silicone counterparts, superior mechanical properties are possessed by PU elastomers. The Young's modulus for them is in the range of 1 to 20 MPa. For synthetic polymer chemists, significant options of tailoring the structures are offered by the composition of PU elastomers; this is useful for meeting specific mechanical requirements. They perform well in humid conditions, under cyclic loading, and in a variety of temperatures, this makes them suitable for implants, i.e., vascular grafts. The blood flow's pulsatile nature under wet-37 C conditions makes these grafts subjected to cyclic loading. Strain rate-dependent viscoelastic behavior is also possessed by the PU-based elastomers.

8.5. INTELLIGENT BIO-MATERIALS

In drug delivery systems, the major achievements have been in the design/development of biomaterials in recent years. At the molecular level, for these special applications, the design of biomaterials is done for matching the functionalities of molecules. A prominent example of such biomaterial is polyrotaxanes. It can be designed in a desirable way for effecting dynamic molecular functions that are similar to the ones of natural tissues via the cyclic compound movement along the linear chain of the polymer. In drug delivery systems, a lot of other elastomeric blends are also being used. New designs of combined and integrated material systems are included in these that are designed for drug delivery in the body and also for drug-coated stents. These are used widely to prevent restenosis in vascular surgery (Dibra et al., 2005). In medical devices, the growing need for lubricious surfaces and coatings has resulted in the manufacture and design of intelligent elastomeric blends. Smooth coatings help physicians in easy maneuvering of medical devices into the regions of delicate tissues and small blood vessels without damaging or injuring them. This has resulted in providing ease to surgeons for performing procedures that were considered impossible in past times. This coating technology is also being used for orthopedic implants with tissue-engineered scaffoldings, bone growth enhancers, and intensive wound care dressings.

8.6. OVERVIEW OF PRODUCTS BASED ON ELASTOMER BIOMEDICAL APPLICATIONS

An overview of elastomer composites and blends that are being used in biomedical applications is provided in this section. Still, a lot of research regarding the development phase of these materials is being carried out. Therefore, discussing the whole range of these materials is not possible here. We are discussing only some of the elastomeric blends that are in use nowadays for biomedical applications.

For biomedical applications, mostly, the elastomers used are formed by synthetic polymers. Thermoset and thermoplastic polymers are included in these. From a clinical and biological standpoint, remodeling capacities and biocompatibility should be exhibited by these elastomers, these properties can be compared with natural extracellular matrix proteins. As far as polymer perspective is concerned, the elastomers should synthesize in the form of large batches and tunable properties should be exhibited. It can be realized by altering polymer processing conditions and synthetic schemes. Some general design principles should be satisfied by the elastomers, these include; (1) non-toxic monomers usage should be allowed that can be excreted or metabolized by the host; (2) incorporation of ester bonds should be allowed for promoting degradation by enzymatic and hydrolysis activity; and (3) chemical crosslinking should be allowed for achieving tunable elastomeric mechanical properties.

8.7. THERMOPLASTIC COMPOSITES AND ELASTOMERS

In medical applications, the largest group of plastics and elastomers in use is formed by thermoplastic polymers. In their preparation, multi-block polymers are utilized in their general synthetic strategy. These polymers are comprised of soft and hard segments. They self-assemble themselves, forming a physically crosslinked network]. Biodegradation capabilities are because of their polyester segments inclusion, within the polymer blocks, contrary to them, the elastomeric properties are because of the physical crosslinks. For medical applications, the commonly used TPEs include polyhydroxyalkanoates, PE, PUs, silicone, and wide photo cross link-able elastomers range such as coumarin-functionalized polyesters, poly(e-caprolactone)-based, and poly(glycerol-co-sebacate)-acrylate (Nwankire et al., 2009; Holvoet et al., 2010).

8.7.1. Polyurethane Composites and Elastomers

Any polymer that consists of an organic unit's chain that is joined through urethane (carbamate) links is called PU. Step growth polymerization forms these PU polymers by reacting a monomer that at a minimum, contains two isocyanate functional groups in combination to another monomer that in catalyst presence, at minimum contains two hydroxyls (alcohol) group (Figure 8.1). Hence, PU elastomers preparation involves three components that are a chain extender, a diisocyanate, and a macrodiol (Ademovic et al., 2005). The color-fastness of a PU elastomer, its high temperature, its resistance to hydrolysis, and its crosslinking behavior, is determined on the basis of the used diisocyanate's nature. On product properties, There are fundamental consequences of macrodial on the properties of the product. Generally, the macrodiol that are used are either diol and ester diol. Oil resistance and outstanding mechanical strength are two primary characteristics of ester-based PUs while excellent resistance to hydrolysis, through bases or acids or hot water, is exhibited by ether-based PUs (Knoerr and HHomann, 2001; Recker, 2001).

Based on the diisocyanate component's nature in their formulation, we designate them either as aliphatic or aromatic. Both of these PUs have similar properties. This makes them excellent materials for medical devices. Properties such as good hydrolytic stability, good biocompatibility, high tensile strength are possessed by them. They can be sterilized via gamma irradiation or ethylene oxide gas, and at low temperatures, they are capable of retaining elastomeric properties (Liua et al., 2006).

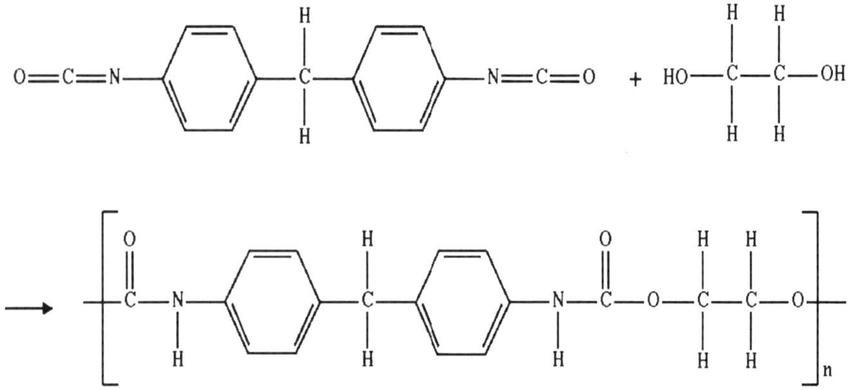

Figure 8.1. Polyurethane polymerization (showed on bottom) from a hydroxyl group (showed on top right) and diisocyanate (showed on the top left).

In the 1930s, PUs were reportedly discovered. However, until the 1950s, they were not been used in biomedical applications. But, today, they are widely used biomaterials. Their applications are vast, spreading all over medical devices, they have their utility in the blood bags and cardiac-assist pumps, and also in chronic implants, i.e., vascular grafts and heart valves. In 1967, Boretos and Pierce first proposed them for use as biomaterials (Boretos and Pierce, 1967). Their blood compatibility and superior mechanical properties have favored their development and use as biomaterials, especially in implanted devices (Lamba et al., 1997). They are vitally used in the manufacture of artificial or prosthetic heart valves.

Even in recent times, artificial heart valves were considered as a mechanical type and was constructed from several combinations of materials such as polymers and metal (Figure 8.2, bottom left). Unidirectional blood flow via mechanical closure of a tilting or pivoting valve is allowed through them. Recently, PU elastomers like biochemically inert synthetic polymers have helped in the formation of prosthetic heart valves (Figure 8.2, bottom right). A wide set of problems (including material fatigue) have been overcome by using polymers in the manufacture of these devices. Not only this, but it has also helped in maintaining functional characteristics and natural hemodynamics of heart valves (Lakshmi et al., 2009). These valves possess the ability for closely simulating or maintaining the natural body hemodynamics because of their soft texture and flexible nature which helps in simulating the lubricity that natural heart valves exhibit, and are able to freely expand and contract (during the blood flow) (Bloomfield, 2002). Their high resistance to fatigue, abrasion, and tearing of PU elastomers sets them best for being used in the prosthetic heart valves (Wiggins et al., 2003).

8.7.2. Polyethylene Composites and Elastomers

A thermoplastic polymer named PE, which is a cross-linked polymer consists of long chains that are produced through a combination of ingredient monomer ethylene (Figure 8.3). We can categorize them into various categories. These categories are based on their branching and density. PE's mechanical properties remarkably depend on variables like type and extent of branching, the molecular weight, and the crystal structure. In biomedical applications, UHMWPE, Ultra-high molecular weight PE, sometimes known as high-performance polyethylene (HPPE) or high-modulus polyethylene (HMPE), is used widely as PE. It consists of long chains having a molecular weight from 2 to 6 million. The load is more effectively transferred through a longer chain. By strengthening intermolecular interactions, this load is transferred to the polymer backbone, which results in high impact strength and toughness. To corrosive chemicals, it is highly resistant, having the exception that of oxidizing acids, low moisture absorption, is efficiently resistant to abrasion, and has a low coefficient of friction. As far as its biocompatibility is concerned, it is nontoxic. In knee and hip replacements, UHMWPE is widely used, and now even for spine implants (Kurtz, 2004).

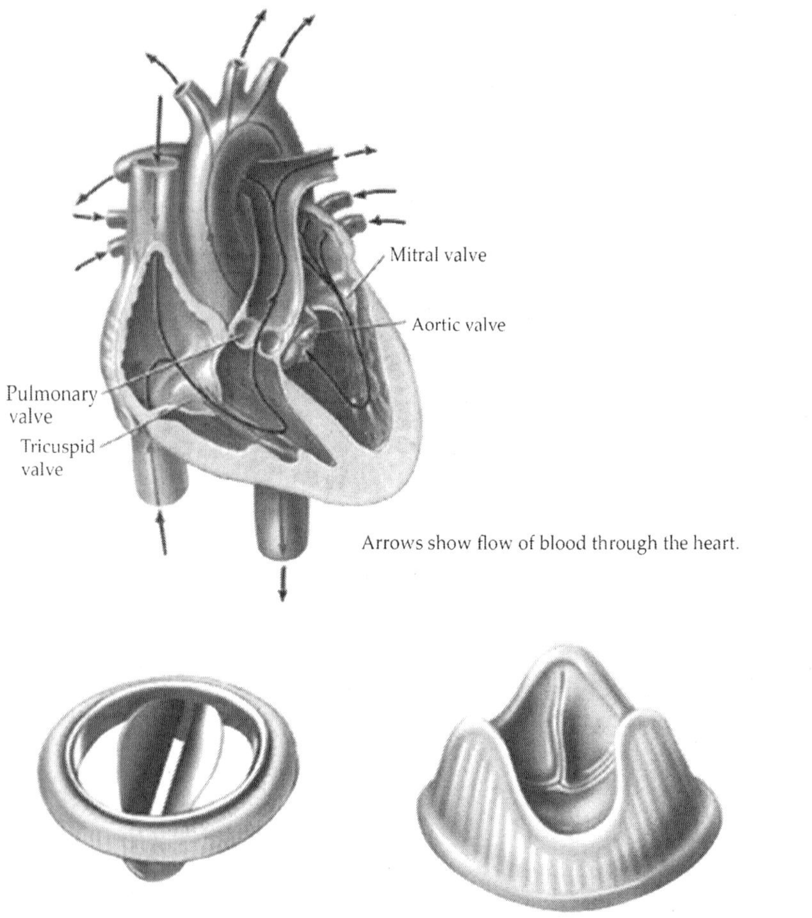

Figure 8.2. At the top, it is an example of artificial heart valves: at the bottom left, the heart demonstrating heart valves positions: i.e., mechanical heart valve is shown, and at the bottom right: polymeric heart valve is shown.

Source: https://www.academia.edu/24417561/Bio_Medical_Applications_of_Elastomeric_Blends_Composites.

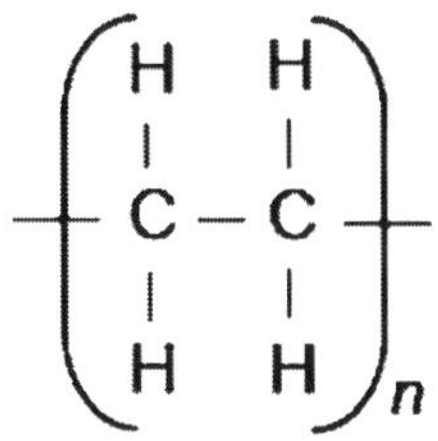

Figure 8.3. Polyethylene polymerization from its ethylene monomer.

Source: https://www.aiche.org/resources/publications/cep/2015/september/making-plastics-monomer-polymer.

UHMWPE's high strength, abrasion resistance, and toughness make it suitable for these applications. Sir John Charnley, in 1962, first used UHMWPE for a medical application and since then, it has dominantly emerged as the bearing material for knee and hip replacements. UHMWPE is being used for making elastomeric composite membranes, as shown by Teoh et al. (1999). His work fabricated a composite membrane of polyether PU and UHMWPE. He combined PU's excellent bio durability and biocompatibility properties with PE's high tensile strength. The resulted product has Young's modulus almost 50 times and the tensile strength almost five times of PU.

An illustration of hip-joint replacement prosthesis is shown in Figure 8.4. Three different kinds of hip replacement implants are being used, i.e., ceramic-on-ceramic, metal-on-metal, and metal and plastic. Ceramic-on-ceramic and metal-on-metal have superior wear rates. For a metal-on-metal implant, some concerns regarding the generated wear debris are present that with time, the body absorbs them. For ceramic implants, the major concern is that there exists a possibility of breakage, i.e., they may break inside the body. The commonly used hip replacement implants are of metal and plastic. As in Figure 8.4, the femoral stem and ball are made through metal (including stainless steel, cobalt chrome, and titanium), while PE is used for making an acetabular cup. For overcoming the PE's high wear rates, UHMWPE is now being used for manufacturing the acetabular cup.

Either by cemented or press-fitting into place, the implant is fixed to the bone. The implant is fixed into the bone snugly when press-fitting and the new bone forms around the implant, securing it in position. While for the cemented method, a special bone fixation agent or cement is used for securing the prosthesis in position. It is illustrated in Figure 8.4.

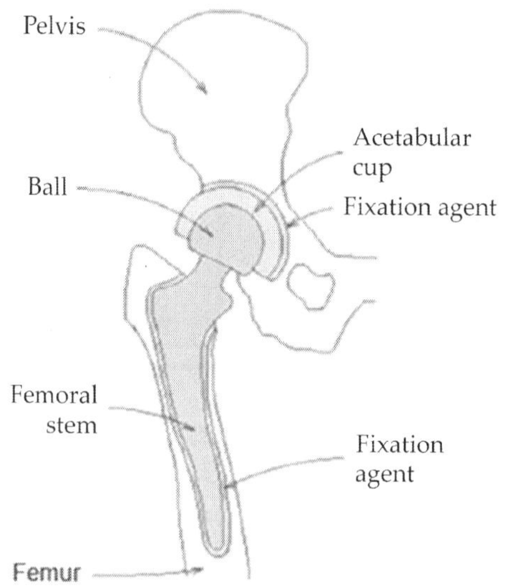

Figure 8.4. Hip joint replacement prosthesis' schematic.

Source: https://www.azom.com/article.aspx?ArticleID=2985.

8.7.3. Silicone Based Elastomers

Silicones are a group of synthetic polymers that are made through the repetition of silicon to oxygen (O_2) bonds, at the same time, silicon is also attached with organic groups (usually methyl groups). Figure 8.5 illustrates it. The repeating unit is referred to as "siloxane." Methyl groups can be substituted by other groups such as trifluoropropyl, vinyl, and phenyl. The presence of both "inorganic" backbone and "organic" groups provide them a set of unique properties. Due to this combination of properties, silicones can be used as compounds, fluids, elastomers, resins, and emulsions. The common type of silicones are polydimethylsiloxanes (PDMS) trimethylsilyloxy terminated (Figure 8.5). Silicones usually incorporate a filler that helps in reinforcing the cross-linked matrix and hence they are regarded as cross-linked polymers (Cui et al., 2010).

For most of the applications, silicone elastomer's strength is generally unsatisfactory without the filler (Noll, 1968). Silicone elastomers in biomedical applications are commonly used as shunts, drains, and catheters. Silicone extrusions fabricated medical devices and the ones that are fabricated with nonsilicone substrates (silicone-coated) for minimizing host reaction, are included in them. The Silastic Foley urology catheters are an example of these devices. They are latex catheters whose interior and exterior are coated by silicone elastomer (Barrett et al., 2010). In extracorporeal machines, silicone membranes and tubing are also widely used. This is mainly because of their blood permeability and compatibility properties. Studies indicate that in several respects, platinum-cured silicone tubing has more potential, which gives it superiority over polyvinyl chloride (PVC). Silicone elastomers are also widely used in reconstructive and esthetic plastic surgery, which includes implants for buttocks, breast, cheek, chin, calf, nose, and scrotum. The silicone breast implant is the most prominent among these implants, and it's usage dates back to 1963. It is still widely popular (John and Foster, 2007; Harmand and Briquet, 1999).

Silicone elastomers exhibit excellent biocompatibility, but in contrast to PU, their mechanical properties are inferior to them. This makes them less favorable for those applications that have a requirement of high mechanical loading, e.g., vascular grafts and heart valves. However, still silicone elastomers have vast medical applications especially urinary catheters and breast implants. Silicone composites or plastics are also used in finger prosthesis implants. An example is shown in Figure 8.6 (Nijst, 2007; Bettingera et al., 2008).

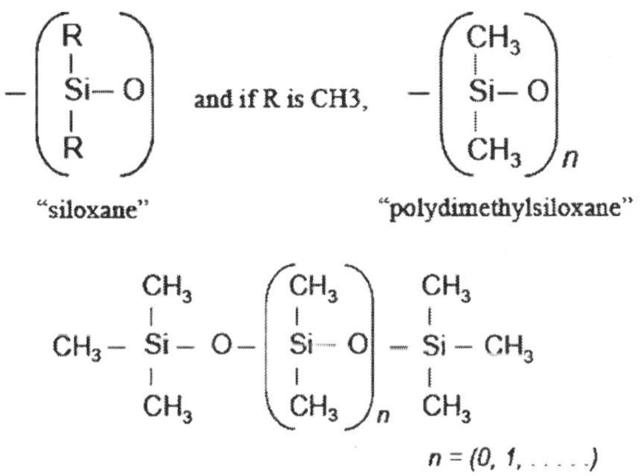

Figure 8.5. Silicone synthesis (shown at bottom) from its monomer (called polydimethylsiloxane).

Source: https://essentialchemicalindustry.org/polymers/silicones.html.

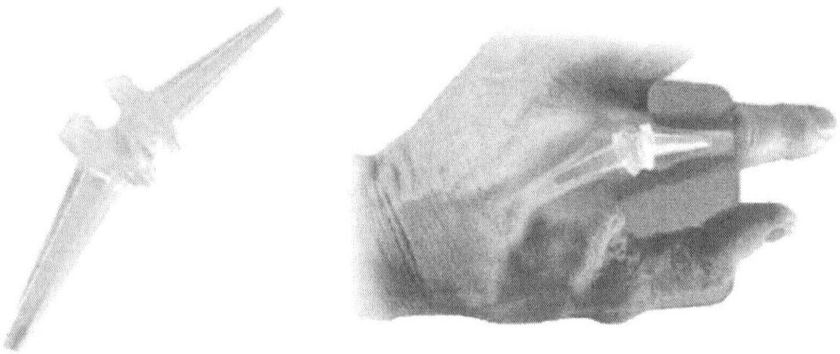

Figure 8.6. A finger prosthesis implant at left and at right shows when it is implanted on hand.

Source: http://www.fingerreplacement.com.

REFERENCES

1. Ademovic, Z., Wei, J., Winther-Jensen, B., Hou, X., & Kingshott, P., (2005). Surface modification of pet films using pulsed ac plasma polymerization aimed at preventing protein adsorption. *Plasma Process. Polym., 2*, 5363.
2. Barr, S., & Bayat, A., (2009). Current implant surface technology: An examination of their nanostructure and their influence on fibroblast alignment and biocompatibility. *Eplasty, 9*, e22.
3. Barr, S., Hill, E., & Bayat, A., (2010). Patterning of novel breast implant surfaces by enhancing silicone bio-compatibility, using biomimetic topographies. *Eplasty, 10*, 246–268.
4. Barrett, D. G., Luo, W., & Yousaf, M. N., (2010). Aliphatic polyester elastomers derived from erythritol and a, x-diacids. *Polym. Chem., 1*, 296–302.
5. Bettinger, C. J., (2011). Biodegradable elastomers for tissue engineering and cell biomaterial interactions. *Macromol. Biosci.* doi: 10.1002/mabi.201000397.
6. Bettingera, C. J., Bruggemanb, P., Borensteinc, J. T., & Langerb, R. S., (2008). Amino alcohol-based degradable poly(ester amide) elastomers. *Biomaterials, 29*(15), 2315–2325.
7. Bloomfield, P., (2002). Choice of heart valve prosthesis. *Heart, 87*(6), 583–589.
8. Boretos, J. W., & Pierce, S. W., (1967). Segmented polyurethane: A new elastomer for biomedical applications. *Science, 158*, 1481–1482.
9. Brandon, H. J., Young, V. L., & Jerina, K. L., (2001). Variability in the properties of silicone gel breast implants. *Plast. Reconstr. Surg., 108*(3), 647–655.
10. Christenson, E. M., Wiggins, M. J., Anderson, J. M., & Hiltner, A., (2005). Surface modification of poly(ether urethane urea with modified dehydroepiandrosterone for improved in vivo biostability). *J. Biomed. Mater. Res., 73A*, 108–115.
11. Colas, A., & Curtis, J., (2009). Biomaterials science. *High Molecular Weight Polyethylene in Total Joint Replacement and Medical Devices.* Academic Press, Elsevier.
12. Cui, W., Zhou, Y., & Chang, J., (2010). Electrospun nanofibrous materials for tissue engineering and drug delivery. *Sci. Technol. Adv. Mater., 11*, 22. doi: 10.1088/1468-6996/11/1/014108.

13. Dibra, A., Kastrati, A., & Mehilli, J., (2005). Paclitaxel-eluting or sirolimus-eluting stents to prevent restenosis in diabetic patients. *N. Engl. J. Med., 353*, 663–670.
14. Gogolewski, S., (1991). *In vitro* and *in vivo* molecular stability of medical polyurethanes: A review. *Trends Polym. Sci., 1*, 47–61.
15. Gunatillake, P. A., Martin, D. J., & Meijs, G. F., (2003). Designing biostable polyurethane elastomers for biomedical implants. *Aust. J. Chem., 56*, 545–557.
16. Harmand, M. F., & Briquet, F., (1999). In-vitro comparative evaluation under static conditions of the hemocompatibility of four types of tubing of cardiopulmonary bypass. *Biomaterials, 20*(17), 1561.
17. Holvoet, S., Chevallier, P., Turgeon, S., & Mantovani, D., (2010). Toward high-performance coatings for biomedical devices: Study on plasma-deposited fluorocarbon films and aging in PBS. *Materials, 3*, 1515–1532.
18. John, L., & Foster, R., (2007). Biosynthesis, properties and potential of natural-synthetic hybrids of poly-hydroxy alkanoates and polyethylene glycols. *Appl. Microbiol. Biotechnol., 75*, 1241–1247.
19. Kang, J., Erdodi, G., & Brendel, M. C., (2010). Polyisobutylene-based polyurethanes. v. oxidative-hydrolytic stability and biocompatibility. *J. Polym. Sci. Part A: Polym. Chem., 48*(10), 2194–2203.
20. Kanyanta, V., & Ivankovic, A., (2010). Mechanical characterization of polyurethane elastomer for biomedical applications. *J. Mech. Behav. Biomater., 3*, 51–62.
21. Kanyanta, V., (2009). *Towards Early Diagnosis of Atherosclerosis-Accurate Prediction of Wall Shear Stress*. PhD thesis, University College Dublin, Ireland.
22. Khorasani, M. T., Zaghiyan, M., & Mirzadeh, H., (2005). Ultra-high molecular weight polyethylene and polydimethylsiloxane blend as acetabular cup material. *Colloids Surf. B, 41*, 169–174.
23. Knoerr, K., &, & HHomann, U., (2001). *Millable Polyurethane Elastomers, Hand-Book of Elastomers* (2nd edn., Vol. 1, pp. 1–25) Marcel Decker, Inc., New York.
24. Koo, N., (2008). The fabrication of a flexible mold for high resolution soft ultraviolet nano-imprint lithography. *Nanotechnology, 19*, 1–4.
25. Kurtz, S. M., (2004). *The UHMWPE Handbook* (Vol. 1, pp. 1–26). Academic Press, New York.

26. Kurtz, S. M., (2009). *UHMWPE Biomaterials Handbook-Ultra-High Molecular Weight Polyethylene in Total Joint Replacement and Medical Devices* (2nd edn., p. 543), Elsevier. ISBN: 978-0-12-374721-1.
27. Lakshmi, P. D., Helene, A. S., Philippe, S., & Ajit, P. Y., (2009). Fluid mechanics of artificial heart valves. *Clin. Exp. Pharmacol. Physiol., 36*(2), 225–237.
28. Lamba, N. M. K., Woodhouse, K. A., & Cooper, S. L., (1997). *Polyurethanes in Biomedical Applications* (Vol. 1, pp. 1–35). CRC Press, Boca Raton.
29. Lelah, M. D., & Cooper, S. L., (1986). *Polyurethanes in Medicine* (Vol. 1, pp. 1–22). CRC Press, Boca Raton.
30. Liua, X., Wona, Y., & Ma, P., (2006). Porogen-induced surface modification of nanofibrous poly(lactic acid) scaffolds for tissue engineering. *Biomaterials, 27*(21), 3980–3987.
31. Lysaght, M. J., & O'Loughlin, J. A., (2000). Demographic scope and economic magnitude of contemporary organ replacement therapies. *ASAIO J., 46*(5), 515–521.
32. Nijst, C. L., Bruggeman, J. P., Karp, J. M., Ferreira, L., & Zumbuehl, A., (2007). Synthesis and characterization of photocurable elastomers from poly(glycerol-cosebacate). *Biomacromolecules, 8*, 3067–3073.
33. Noll, W., (1968). *Chemistry and Technology of Silicones* (Vol. 1, pp. 1–25). Academic Press, New York.
34. Nwankire, C. E., Ardhaoui, M., & Dowling, D. P., (2009). The effect of plasma-polymerized silicon hydride-rich polyhydrogenmethylsiloxane on the adhesion of silicone elastomers. *Polym. Int., 58*(9), 996–1001.
35. Nwankire, C. E., ONeill, L., Byrne, G., & Dowling, D. P., (2008) The effect of plasma polymerized si-rich polymethyl hydrogen siloxane (PHMS) on the adhesion of silicone elastomer. In: *Proceedings of the 31st Annual Meeting of the Adhesion Society* (p. 436).
36. Onatea, J. I., Cominb, M., Bracerasa, I., (2001). Wear reduction effect on ultra-high-molecular-weight polyethylene by application of hard coatings and ion implantation on cobalt-chromium alloy, as measured in a knee wear simulation machine. *Surf. Coat. Technol., 142–144*, 1056–1062.
37. Ratner, B. D., (2007). The catastrophe revisited: Blood compatibility in the 21st century. *Biomaterials, 28*(34), 5144–5147.

38. Recker, K., (2001). *Cast Polyurethane Elastomers, Handbook of Elastomers*, (2nd edn., Vol. 1, pp. 1–22). Marcel Decker, Inc., New York.
39. Ripple, W. S., & Simons, J., (2007). *Thermoplastic Elastomers in Medical Devices*. Technical contribution for Med Plast Supplement.
40. Shanshan, L., Daniel, M. D., & Yi, C., (2010). Designed biomaterials to mimic the mechanical properties of muscles. *Nature, 465*(7294), 69–73.
41. Stokes, K., & McVenes, R., (1995). Polyurethane elastomer biostability. *J. Biomater. Appl., 9*, 321–354.
42. Szycher, M., & Reed, A., (1992). Biostable polyurethane elastomers. *Med. Device Technol., 3*, 42–51.
43. Szycher, M., (1998). Polyurethanes in vascular grafts. *Elastomer World, 218*, 44.
44. Teoh, S. H., Tang, Z. G., & Ramakrishna, S., (1999). Development of thin elastomeric composite membranes for biomedical applications. *J. Mater. Sci. Mater. Med., 10*(6), 343–352.
45. Tilak, M. S., (2001). *Dip Molding of Polyurethane and Silicone for Latex-Free, Nonallergic Products.* Medical device and diagnostic Industry.
46. Wiggins, M. J., Anderson, J. M., & Hiltner, A., (2003). Effect of strain and strain rate on fatigue accelerated biodegradation of polyurethane. *Biomed. Mater. Res., 66A*, 463–475.

INDEX

A

Abrasion 2, 14, 21, 30, 38
Aircraft tires 230
Air-dried sheets (ADS) 56
Ample 209
Atmospheric pressures 236
Atomic force microscopy (AFM) 117
Automotive passenger tires 230

B

Biodegradation capabilities 251
Boltzmann factor 122
Bow dome 226, 227, 228
Brittle plastic 2
Bromobutyl rubbers (BIIR) 53
Butyl Rubber (IIR) 15

C

Carbon black (CB) 10, 127
Carbon nanotubes (CNTs) 33
Carboxylated Nitrile Butadiene Rubber (XNBR) 55
Chemical bond stretching 3
Chemical resistance 2, 14, 18, 26, 30, 38
Chemical vapor deposition (CVD) 182
Chlorination 15
Chlorobutyl rubbers (CIIR) 53
Chloroprene 76, 104
Chlorosulfonated Polyethylene (CSM) 55
Chlorosulfonated polyethylene (CSPE) 142, 237
Cinder 56
Coating technology 250
Combat Rubber Raiding Craft (CRRC) 237
Cosmetic surgery 246
Cryo-ultramicrotomy 134
Crystallinity constraints 2
Crystal structure deformation 3

D

Dehydrochlorination 15
Dental implants 246
Dirt 56
Drug delivery systems 250

E

Earth's surface 236
Elasticity 52, 53, 55, 65, 69, 72, 73, 74, 79, 84, 90, 93, 97, 103, 107
Elastic strain 3
Elastomeric blends 246, 247, 249, 250, 251

Elastomeric polymer light-emitting device (EPLED) 185
Elastomers 2, 4, 14, 15, 16, 17, 18, 19, 22, 24, 26, 27, 30, 31, 32, 38, 41, 43, 47
Enthalpy change 118
Equilibrium 168
Ethylene Propylene Diene Monomer (EPDM) 15
Ethylene propylene diene monomer (EPDM) rubber 128
Ethylene-vinyl acetate 52, 105, 111, 112
Ethylene-Vinyl Acetate Copolymer (EVA) 55
External quantum efficiency (EQE) 185

F

Fabrication 178, 198, 199, 201
Fluorine 2, 26
Fluorocarbon elastomers (FKM) 26
Fluorosilicones 86
Full width half maximum (FWHM) 169

G

Gibbs free energy 118
Graphene 209, 210, 217, 219, 220, 221, 222
Graphene oxide (GO) 209
Graphite 209, 210, 215, 217, 219, 220, 221, 222
Graphite intercalation compounds 209
Graphite nanoplatelets (GNPs) 209

H

Heat 52, 53, 54, 55, 57, 63, 68, 70, 72, 74, 79, 84, 85, 89, 90, 92, 101, 108
High-modulus polyethylene (HMPE) 253
High-performance polyethylene (HPPE) 253
Hooke's Law 2
Hydrogenated nitrile rubbers (HNBR) 71
Hysteresis 8

I

Indium tin oxide (ITO) 194
Inorganic materials 247, 248
Intermolecular interactions 121, 127
Interpenetrating polymer network (IPN) 124
Isobutylene-Isoprene Rubber (IIR) 67
Iso-dimensional nanofillers 208
Isomerization 15
Isoprene monomers 58

K

Kinetics 2, 10, 11, 41, 42, 43, 44, 49

L

Light-emitting diodes (LEDs) 180
Lower critical solution temperature (LCST) 118

M

Mechanical stretchability 178, 193
Medical devices 246, 247, 249, 250, 252, 256

Melt Processible Rubbers (MPR) 95
Microscopic imaging 133
Military hovercraft 238
Mobility 208
Moisture condensation 229
Molecular structure 12, 38
Multiwall carbon nanotubes (MWCNTs) 210, 213

N

Nanogenerator (NG) 196
Natural rubber (NR) 2
Nitrile rubbers 70
Nitrogen 2, 56, 113

O

Oil-extended natural rubber (OENR) 58
Optical microscopy (OM) 117
Oxidizing acids 253
Oxygen (O2) 2
Ozone 52, 53, 54, 55, 58, 60, 62, 65, 66, 68, 69, 70, 73, 74, 77, 79, 82, 84, 86, 88, 89, 91, 92, 94, 98, 100

P

Piezoelectric elastic composite (PEC) 196
Poly(3,4-ethylene dioxythiophene (PEDOT) 178
Polyacetylene (PA) 178
Polyacrylic rubber 52
Polyaniline (PANI) 178
Polycarbonate esteramides (PCEAs) 30
Polydimethylsiloxane (PDMS) 27
Poly(dimethylsiloxane) (PDMS) rubbers 211
Polyether block amides 52
Polyethyleneimine (PEI) 186
Polyisobutylene 14
Polymer blends 35
Polymers 2, 4, 5, 11, 19, 20, 21, 27, 29, 31, 36, 38, 39, 41, 46
Polypropylene 2, 32, 47
Polypyrrole (PPY) 178
Polysiloxanes 14
Polysulfide (PSR) 18
Polysulfide rubbers (PSR) 87
Polyurethane (PU) elastomer 8
Polyurethanes (PU) 79
Polyvinyl chloride (PVC) 33
Polyvinylidene fluoride (PVDF) 196
Power conversion efficiency (PCE) 194
Premature failure 246

R

Radiation 52, 62, 66, 74, 89, 92
Reinforcement fillers 86
Resistance temperature detector (RTD) 192
Ribbed smoked sheets (RSS) 56
Rubber 52, 53, 54, 56, 57, 58, 60, 63, 64, 66, 67, 68, 69, 70, 71, 73, 76, 80, 85, 86, 89, 90, 93, 95, 96, 99, 103, 104, 105, 106, 107, 108, 109, 110, 111, 112, 113
Rubber-coated fabric 234
Rubber-like elasticity 208
Rubber-like materials 208
Rubber's molecular orientation 170
Rubbery polymers 119

S

Scanning electron microscopy (SEM) 117
Silicon 2, 27
Silicone breast implant 257
Silicone rubber (SR) 52
Single-walled CNTs (SWCNTs) 182
Skincare 246
Sonar Rubber Domes 225
Sound energy 226
Space blimp 235, 236
Strain energy 162
Strain sensors 188
Stretchability 178, 179, 180, 182, 183, 187, 188, 189, 191, 192, 193, 194, 195, 196, 197, 198, 202, 204
Stretchable light-emitting systems 185
Stretchable technology 178, 198
Styrene-Butadiene Rubber (SBR) 14
Styrene-butadiene-styrene (SBS) 29
Styrene/ethylene-butylene copolymer (SEBS) 93
Styrene/isoprene copolymer (SIS) 93
Styrene-isoprene-styrene (SIS) 29
Styrenic block copolymers 29, 30
Sulfur vulcanization 52
Synthetic Rubber 52, 53, 108

T

Thermodynamics 2, 5, 7, 10, 41, 44, 45, 50
Thermoplastic 2, 3, 21, 28, 30, 31, 32, 33, 41, 43, 45, 46, 47, 49
Thermoplastic elastomers (TPEs) 2
Thermoplastic Olefin Elastomers (TPO) 95
Thermoplastic vulcanizates (TPV) 95
TPOs (Polyolefin thermoplastic elastomers) 31
Transmission electron microscopy (TEM) 117
Trans polyisoprene (TPI) 138
Treadwear 232, 233, 243

U

Upper critical solution temperature (UCST) 118

V

Vascular surgery 250
Volatile elements 56
Vulcanized rubber 52, 96

Z

Ziegler-Natta mixtures 61